FUZZY TOPSIS

FUZZY TOPSIS
Logic, Approaches, and Case Studies

Mohamed El Alaoui

CRC Press
Taylor & Francis Group
Boca Raton London New York

CRC Press is an imprint of the
Taylor & Francis Group, an **Informa** business

MATLAB® is a trademark of The MathWorks, Inc., and is used with permission. The MathWorks does not warrant the accuracy of the text or exercises in this book. This book's use or discussion of MATLAB® software or related products does not constitute endorsement or sponsorship by The MathWorks of a particular pedagogical approach or particular use of the MATLAB® software.

First edition published 2021
by CRC Press
6000 Broken Sound Parkway NW, Suite 300, Boca Raton, FL 33487-2742

and by CRC Press
2 Park Square, Milton Park, Abingdon, Oxon, OX14 4RN

© 2021 Mohamed El Alaoui

CRC Press is an imprint of Taylor & Francis Group, LLC

The right of Mohamed El Alaoui to be identified as author of this work has been asserted by him in accordance with sections 77 and 78 of the Copyright, Designs and Patents Act 1988.

Reasonable efforts have been made to publish reliable data and information, but the author and publisher cannot assume responsibility for the validity of all materials or the consequences of their use. The authors and publishers have attempted to trace the copyright holders of all material reproduced in this publication and apologize to copyright holders if permission to publish in this form has not been obtained. If any copyright material has not been acknowledged please write and let us know so we may rectify in any future reprint.

Except as permitted under U.S. Copyright Law, no part of this book may be reprinted, reproduced, transmitted, or utilized in any form by any electronic, mechanical, or other means, now known or hereafter invented, including photocopying, microfilming, and recording, or in any information storage or retrieval system, without written permission from the publishers.

For permission to photocopy or use material electronically from this work, access www.copyright.com or contact the Copyright Clearance Center, Inc. (CCC), 222 Rosewood Drive, Danvers, MA 01923, 978-750-8400. For works that are not available on CCC please contact mpkbookspermissions@tandf.co.uk

Trademark notice: Product or corporate names may be trademarks or registered trademarks and are used only for identification and explanation without intent to infringe.

Library of Congress Cataloging-in-Publication Data
Names: El Alaoui, Mohamed, author.
Title: Fuzzy TOPSIS : logic, approaches, and case studies / Mohamed El Alaoui.
Other titles: Fuzzy technique for order of preference by similarity to ideal solution
Description: First edition. | Boca Raton : CRC Press, 2021. | Includes bibliographical references and index.
Identifiers: LCCN 2020052970 (print) | LCCN 2020052971 (ebook) | ISBN 9780367767488 (hbk) | ISBN 9781003168416 (ebk)
Subjects: LCSH: Decision support systems. | Fuzzy systems.
Classification: LCC T58.62 .E43 2021 (print) | LCC T58.62 (ebook) | DDC 658.4/030285633--dc23
LC record available at https://lccn.loc.gov/2020052970
LC ebook record available at https://lccn.loc.gov/2020052971

ISBN: 978-0-367-76748-8 (hbk)
ISBN: 978-0-367-76749-5 (pbk)
ISBN: 978-1-003-16841-6 (ebk)

Typeset in Times
by MPS Limited, Dehradun

Contents

Preface vii
Author ix
Introduction xi

Chapter 1 Uncertainty 1

Chapter 2 Nonclassical Logics 15

Chapter 3 Fuzzy Logic 31

Chapter 4 Frequently Used Multicriteria Decision-Making Methods 41

Chapter 5 TOPSIS Methodology and Limits 65

Chapter 6 Fuzzy TOPSIS 95

Chapter 7 Intuitionistic Fuzzy TOPSIS 135

Chapter 8 Other Fuzzy TOPSIS Approaches 175

Index 199

Preface

Decision making consists of choosing the best alternative according to several criteria. But the decision-making process is uncertain on several levels. First, the decisions taken are relative to future events that can only be predicted. Second, the decision makers responsible for making these predictions, no matter how cautious they may be, will not be able to proclaim perfect knowledge, exempt from errors and misjudgments. To these we can add issues of information availability and relevance.

Due to the historical development of uncertainty theories, probability was considered for a long period to be a synonym of uncertainty. Recently, fuzzy logic has imposed itself—especially in decision making—dogmatically. However, overuse makes it appear absurd, especially to novices. That is why Chapter 1 distinguishes the causes, sources, and types of uncertainty. To which we have to add, according to Zimmermann, the context, kind of available inputs, desired outputs, and inference mechanism used in order to choose the appropriate model to represent uncertainty. Then, without claiming to provide the magic tool, fuzzy logic is reviewed as a logic in Chapter 2, completing and extending classical and multivalued logics, and as a tool to model uncertainty that fits numerous situations described in the first part. The chapter also discusses other unconventional logics, such as quantum theory, and the reasons for their rejection. Chapter 3 reviews the required mathematical tools, differentiates fuzzy sets from classical sets, and discusses some extensions of fuzzy sets. Based on a general decision-making scheme, and the absence of an absolute best method, Chapter 4 is devoted to frequently used decision-making methods (SAW, AHP, ELECTRE, PROMETHEE, and VIKOR).

Chapter 5 starts from the initial methodology of TOPSIS with a relevant MATLAB® algorithm. It reviews the main causes of rank reversal, namely ideal solutions, and normalization, in addition to their combined effect. The chapter also highlights rank reversal in other methods and discusses the impact of the distance used on the final ranking of alternatives.

To capture both quantitative and qualitative criteria, TOPSIS has been extended using fuzzy logic. Chapter 6 reviews some extensions with relevant examples, in addition to a corresponding MATLAB algorithm. Qualitative criteria, as opposed to quantitative ones, depend heavily on the person making the assessments. This is why they require several decision makers to reduce bias. To merge these opinions into a consensual one, a consensus-reaching method is integrated. The chapter also discusses some frequently used algorithm parameters, including aggregation, distance, normalization, and the choice of ideal solutions. Furthermore, it presents TOPSIS adaptations to different types of fuzzy sets.

To adapt to the intuitionistic fuzzy context, Chapter 7 presents aggregation as a part of the information-integration process, discussing the classically used forms in the intuitionistic fuzzy context, and improved ones. It also presents the adapted distances and consensus-reaching process, the latter of which permits a better optimized function with fewer operations. Then a brief review of some intuitionistic fuzzy TOPSIS approaches is presented. After that, the proposed approach for a consensual TOPSIS approach (with examples) is extended for interval-valued and continuous intuitionistic fuzzy sets using trapezoidal representation. The last section discusses a method to derive attribute weights in intuitionistic fuzzy TOPSIS when they are unknown.

Chapter 8 reviews emerging fuzzy TOPSIS approaches using neutrosophic sets, hesitant fuzzy sets, and Pythagorean fuzzy sets.

Author

Mohamed El Alaoui holds a PhD in industrial engineering from ENSAM Meknes, Morocco, and an engineering degree in industrial and production engineering from the same institution. His PhD thesis, titled *Fuzzy Logic and Optimization with Applications in Industrial and Production Engineering*, treated several applications, among them biodiesel, aggregation, consensus, similarity, goal programming, transportation, and partitioning. In his academic career, he has taught various courses related to industrial engineering (project management and quality management) and mechanics (solid mechanics and hydraulics). He has also worked as a corporate trainer. He is a reviewer for leading journals.

Introduction

Since I was a child, I have had great difficulty making decisions, especially when it comes to limited choices from a small number of alternatives. It is true that as we grow up, we become more and more attentive to the realities that surround us and guide our choices. Nevertheless, the difficulties seemed persistent. Miraculously, my thesis treated decision making, and it was only by addressing it more deeply that I realized that the slowness I was accused of in making my choices was simply due to the lack of data related to the problems (criteria, weights ... and so forth) and the lack of theoretical background to model and process it.

Even in the presence of a powerful theoretical artillery, an in-depth knowledge of the problem under study is required. However, there is a tendency to confuse the inherent uncertainty in a subject with a lack of mastery. Thus, Chapter 1 deals with uncertainty, the fuzzy side, trying to answer the following questions: Can we build a model that is free of uncertainty? What are the forms of uncertainty we encounter? Can they be treated uniformly by a single theory?

It is true that this book advocates the use of fuzzy logic to model uncertainty. However, this does not mean at all that it is the miracle theory to solve everything. All the more, it remains poorly understood even by practitioners. All beginners in fuzzy logic who want to apply it to their research fields will face, at least, these major problems:

- Human nature is resistant to change. Even more, fuzzy logic is not a simple knowledge that will be added to or will complement a certain set of information, but questions our understanding. Breaking free from the probabilistic model in order to use the possibilistic one requires recognizing the limits of the former, which we previously thought could model all forms of uncertainty on its own.
- Works applying fuzzy logic only partially use the arsenal it provides, which does not justify its use in the eyes of readers and especially beginners.
- Specialized works in fuzzy logic are often deeply steeped in the theoretical side of the field, invoking logical and mathematical details that are useless to most readers.
- In addition, fuzzy logic and possibility are often presented as generalizations of classical logic and probability, respectively, which must be discussed carefully (Bier, 1990). The result of a rolled die or a tossed coin is perfectly modeled by probability; the introduction of a new theory of uncertainty does not mean that the old one should be thrown away. Similarly, the transitions between classical logic and fuzzy logic are not that obvious. The multiplicity of definitions and uses require careful clarification.

For these different reasons—and as all reasoning is based on the logic adopted—Chapter 2 discusses the limits of classical logic, some nonclassical logics like quantum mechanics, and the reasons for the rejection of these logics. Chapter 3 identifies the necessary fuzzy set theory tools used in this book. I invite readers who might already think that the rest of this book will be a plea in favor of fuzzy logic to a certain epoche, a suspension of judgment (Laos, 2015). The need is all the more persistent in an age when we are being pushed to react instantly to multiple stimuli.

It should be noted that the rest of this book is not a plea for the Technique for Order of Preference by Similarity to Ideal Solution (TOPSIS) method as a miracle solution. Otherwise, how could we explain the plethora of methods that exist in the literature, if it is not to deal with various problems and realities? Thus, Chapter 4 considers classically used and well-known methods, with relevant examples. It also discusses the main steps of the decision-making process.

Chapter 5 reviews the original TOPSIS algorithm with an illustrative example and proposes a relevant MATLAB algorithm. It maintains that like other methods TOPSIS suffers from rank reversal. Thus, the chapter reviews the main causes treated in the literature, namely ideal solutions and normalization. Since none of the proposed approaches that deal with just one of those two aspects totally prevents rank reversal, the chapter discusses five interpretations of ideal solutions and 10 normalizations. It also shows experimentally—without pretending to find the magical solution—that some combinations are more likely to cause rank reversal than others. It also discusses 15 distance metrics and their effects on the final ranking.

Chapter 6 reviews TOPSIS in the fuzzy context. It presents precursor approaches (Chen, 2000; Negi, 1989) with relevant examples and proposes a MATLAB algorithm. It also discusses the frequently used distance metrics, ideal solutions, normalization methods, and aggregation operations in fuzzy TOPSIS. In addition to rank reversal, increasing attention is given to group consensus. In that sense, the chapter also includes a consensus-reaching method adapted to the fuzzy context. Experimental results show that the proposed combination better fits the definition of TOPSIS, because each alternative dominates the next in being both closer to the positive ideal solution and further from the negative ideal solution. The chapter also discusses TOPSIS extensions using type 1 and type 2 fuzzy sets, with related adaptations and examples.

Chapter 7 is devoted to intuitionistic fuzzy TOPSIS. It presents aggregation as a step of information integration and discusses the usual aggregation functions and their categories. Then it presents the frequently used intuitionistic fuzzy aggregation operators, discusses their limits, and presents extensions. It also presents the adapted intuitionistic fuzzy distances and consensus-reaching process. The latter is shown, through an example, to permit a better-optimized function with fewer operations. Then a brief review of some intuitionistic fuzzy TOPSIS approaches is presented. After that, the proposed approach for consensual TOPSIS, with examples, is extended for interval-valued and continuous intuitionistic fuzzy sets using trapezoidal representation. The last section discusses a method to derive attribute weights in intuitionistic fuzzy TOPSIS when they are unknown, with an example.

Chapter 8 is a sort of literature review on Pythagorean fuzzy TOPSIS, neutrosophic TOPSIS, and hesitant fuzzy TOPSIS. It presents the required mathematical background, discusses possible extensions, and proposes a simple fuzzy consensus in each situation, with examples.

As TOPSIS rests on ideal solutions, one could say that some people believe in being at the nadir (worst possible solution) while proclaiming the utopia (best possible solution).

REFERENCES

Bier, V. M. 1990. "Is Fuzzy Set Theory More General than Probability Theory?" In *[1990] Proceedings. First International Symposium on Uncertainty Modeling and Analysis*, 297–301. http://sci-hub.tw/10.1109/ISUMA.1990.151267.

Chen, Chen-Tung. 2000. "Extensions of the TOPSIS for Group Decision-Making Under Fuzzy Environment." *Fuzzy Sets and Systems* 114 (1): 1–9. http://sci-hub.tw/10.1016/S0165-0114(97)00377-1.

Laos, Nicolas. 2015. *The Metaphysics of World Order: A Synthesis of Philosophy, Theology, and Politics*. USA: Wipf and Stock Publishers.

Negi, Devendra Singh. 1989. "Fuzzy Analysis and Optimization." Department of Industrial Engineering: Kansas State University.

1 Uncertainty

AMBIGUITY AND VAGUENESS

"Who speaks of vagueness should himself be vague."

(Russell, 1923)

"It seems that the only certain aspect of science is that it is uncertain."

(Ronen, 1988)

"It is the mark of an educated mind to rest satisfied with the degree of precision which the nature of the subject admits and not to seek exactness where only an approximation is possible."

(Aristotle)

When introducing me to cooking, my mother asked me to add "a little of X." After I put in the first ingredient, she said, "You add also a little of Y." While I was trying to put in approximately the same quantity, she intervened: "Not that much." Trying to be funny, I said, "So a little of X does not equal a little of Y?"

How much of the second ingredient would be too much? The same question can be asked with respect to the lower limit—the amount below which the ingredient used would be insufficient.

It is not because we do not know the definition of "a little" (small amount), nor because we lack balances to weigh quantities down to the gram or even the milligram; this uncertainty is mainly due to the multiplicity of interpretations that can be linked to the situation and the absence of a distinguishable limit that would allow the unambiguous separation of different possible states. According to Peirce's 1902 definition of vagueness in logic,

> ...a proposition is vague when there are states of things concerning which it is intrinsically uncertain whether, had they been contemplated by the speaker, he would have regarded them as excluded or allowed by the proposition. By intrinsically uncertain we mean not uncertain in consequence of any ignorance of the interpreter, but because the speaker's habits of language were indeterminate; so that one day he would regard the proposition as excluding, another as admitting those states of things. Yet this must be understood to have reference to what might be deduced from a perfect knowledge of his state of mind; for it is precisely because these questions never did, or did not frequently, present themselves that his habit remained indeterminate.

(Baldwin, 1902)

Eubulides formulated the sorites paradox (named from the Greek *soros*, meaning "heap"), also known as the heap paradox, as early as the fourth century BC. It can be approached in two ways. In the first, everyone can say that a single grain of sand cannot be a heap. Adding a second grain does not make it a heap either. Continuing in the same way by adding a grain at a time, it is obvious that the result will necessarily be a heap, since there seems to be no valid reason to stop at one point rather than another (Black, 1963). The second approach, the decreasing one, starts with a heap of sand and states that after removing a single grain, the result will still be a heap, and after removing a second, and so on. The question that can be deduced is: How many grains are needed to have a heap?

Other variants also exist, such as baldness, length, age, wealth, and adulthood (see Chapter 2), where any line arbitrarily drawn between bald and hairy, long and short, old and young, rich and poor, major and minor will be challenged. Frege sees this vagueness in the definitions of these words as a defect in human language that must be eliminated (Ebert & Rossberg, 2016). According to Williamson, such a concrete language is an unachievable ideal. If we tried to establish it, it would be based on vague stipulations that would taint the end result (Williamson, 1994). One can simply notice that the phrase "a pile minus a grain is always a pile" involves a qualitative notion ("pile") and a quantitative one ("grain"), whose subtraction is either contradictory or nonmodelable (Borislav, 2018).

From Peirce's point of view, language is and will remain vague. However, this is only harmful if it leaves the borders of the problem to be dealt with too wide. The same vision is shared by Russell (1923), who adds that even though the logical connectors "OR" and "AND" are certainly less vague than other terms of everyday language, they cannot deviate from the rule when they are jointly employed with other vague elements. Russell even attacks scientific units such as the second and the meter. Relevantly, the 26th General Conference on Weights and Measures, held November 13–16, 2018, redefined four of the seven basic units of the International System of Units, knowing that the meter is currently based on Planck's constant, which is measured by two methods—one using a watt balance (also called a Kibble balance) and one involving the Avogadro constant—have provided slightly different results. (It should be noted that this small difference can only affect research, especially in nanotechnology.)

Russell also points to the discrepancy between words and their meanings. As long some words may have several meanings, there is ambiguity in the overall meaning of a statement. While admitting the impossibility of defining a language devoid of ambiguity, Russell advocates the use of semantics in which each word or couple of words has a unique meaning. There are certainly bald people, there are certainly hairy people, and there are people whom you cannot call either. The principle of the excluded middle cannot be applied for vague terms: the words are vague and that will not change. However, words derive their sense from context. Thus, the juxtaposition of vague words can lead to a meaning that is less vague. Russell (1923) concludes, "Vagueness, clearly, is a question of degree, depending on the extent of the possible disparities between the different systems evoked by the same representation. Precision, on the other hand, is an ideal limit."

Uncertainty, which is the generic term (Raskin & Taylor, 2014), can be divided into two main branches (Klir, 1987):

- Ambiguity means the existence of several possible alternatives which are left unspecified. Zhang (1998) points out that an expression is ambiguous if it has several paraphrases that are not paraphrases of one another.
- Vagueness refers to situations where clear boundaries cannot be defined.

While fuzzy sets represent a basic mathematical framework for dealing with vagueness, fuzzy measures are a general framework for dealing with ambiguity (Klir, 1987). A similar classification of fuzzy linear programming has been proposed by Inuiguchi and Ramík (2000).

Before showing the need for fuzzy logic, do we need to model uncertainty? In trying to create a new collaboration with a third-party research team, the obvious question is: What more can everyone bring? That is why I proposed modeling uncertainty based on fuzzy logic. While I expected to justify the choice of fuzzy logic rather than a probabilistic model, the answer was formal: there is no uncertainty in our model. That is why I asked myself, is there an uncertainty free model?

UNCERTAINTY IN EXACT SCIENCES

Avoiding philosophical quarrels on the subject (Sanford, 1995), uncertainty is everywhere: chemistry (Tchougréeff, 2016), economics (Köhn, 2017), even where you do not expect it (Deemter, 2010). While its existence may go unnoticed in management (Nicolai & Dautwiz, 2010) and music (Venrooij & Schmutz, 2018), it can pose ethical problems in justice (Braun, Schickl, & Dabrock, 2018; Keil & Poscher, 2016). In a world where one has the illusion of mastering certain concepts to the last comma, people assume—without proof—that exact sciences, such as mathematics, would not be tainted with uncertainty (Hoyt, 1941). However, the definition of exact science is itself vague.

Far from the noble structure the term "exact science" may imply, it is not based on any universally valid principle (Planck, 1949); since all sciences are based on a number of axioms, which are by definition accepted without demonstration. Can we build science without presuppositions? According to Planck, scientific thought must relate to something. However, no uniformly acceptable worldview has been defined. Hence, it is impossible to place an exact science in a fixed and inclusive content (Planck, 1949). An extensive discussion on the meaning of presuppositions in science is given by Kattsoff (1957).

The term "exact science" has excessively been linked to mathematics (Grant, 2007; Neugebauer, 1969), what pushes the disciples of many fields of thought, including literary ones, try to make their respective specialties as mathematical as possible so that the results obtained can be considered reliable. Working in an inexact discipline can seem pejorative, whereas the appearance of exactitude is perceived as a form of nobility, as in economics (Nemchinov, 1962), biology (Smart, 1959), and meteorology (Sutton, 1954). According to Dompere (2013), the characteristics of an exact science are:

- Exact measurements and quantities
- Observation, experimental testing, and verification

- Sensory data as a material entry into the exact reasoning
- Reliable logical laws, mathematical representation, and mathematics based on exact logic
- Axioms of universal validity with exact symbols and exact methods of analysis and synthesis
- Certainty of logic and conclusion

Other, less idealistic and prominent visions of mathematics also exist. Russell defines mathematics as the subject in which we never know what we are talking about, or whether what we say is true (Jourdain, 1912). The need to introduce some flexibility is mentioned by Einstein: "as far as the laws of mathematics refer to reality, they are not certain; and as far as they are certain, they do not refer to reality" (Berlinghoff, Grant, & Skrien, 2001).

In our quest for precision, we have tried to adapt the real world to mathematical models that make no provision for fuzziness (Kaufmann, 1975). We know nothing about reality other than by our models, our representations, our laws more or less true, and our acceptable approximations. Moreover, the model of something for one person is not exactly the same model of that same phenomenon for others; the formula can remain the same, but the interpretation may be different.

The universe is perceived through the help of models that perfect themselves by embodying each other, at least until a revolution in ideas no longer supports them (Kaufmann, 1975). On the other hand, one can observe the ongoing evolution of what is meant by "exact science" over time (Watson, 1939), implying that the current state is also not final.

An exact description of the world is simply impossible. That is a fact that we have to accept, and it is up to us to adjust to it. As a result, instead of pursuing an unattainable ideal, the problem becomes reducing the inevitable imprecision to a level of relative unimportance. We need to balance the needs for accuracy and simplicity, and reduce complexity without oversimplification in order to respond adequately to each situation (Bellman & Giertz, 1973).

UNCERTAINTY CAUSES

Once we have recognized the undeniable existence of uncertainty and the truth that even the so-called exact sciences are based on approximations (Russell, 1951), the next step is to recognize the causes of uncertainty. As Zimmermann (2011) has pointed out, there is confusion between the type, the source or cause, and the method or model to deal with uncertainty. For historical reasons, which have greatly influenced curriculums, as soon as we hear the word uncertainty, we automatically think of measurement uncertainty, probability, and so on.

The widely used approach to classifying sources of uncertainty divides it to two types (Asselt, 2000; Li, Chen, & Feng, 2013; Sallak, Aguirre, & Schon, 2013):

- Random uncertainty (Henrion & Fischhoff, 1986)—also known as variability (Asselt, 2000), objective uncertainty (Natke & Ben-Haim, 1997), external uncertainty (Kahneman & Tversky, 1982), "random uncertainty" (Henrion &

Fischhoff, 1986), stochastic uncertainty (Helton, 1994), and inherent, irreducible, fundamental, real, or primary uncertainty (Koopmans, 1957)—is due to the inherent variability of natural phenomena. It exists independently of us. The best answer for a coin flip can only be "heads or tails," because we can only predict the result at a probability of 1/2, and it is not by collecting more data that we will achieve the ultimate answer.
- Epistemic uncertainty—also known as knowledge uncertainty (Reduction, National Research Council U S. Committee on Risk-Based Analysis for Flood Damage, 2000), subjective uncertainty (Helton, 1994), internal uncertainty (Kahneman & Tversky, 1982), incompleteness (Schomberg, 1993), functional uncertainty (Natke & Ben-Haim, 1997), and secondary uncertainty (Koopmans, 1957)—is due to humans and their inability to apprehend our environment, whether due to a lack of model measures or something else. In contrast to random uncertainty, epistemic uncertainty can be reduced or eliminated through more data and better methods.

Similarly, Walley (1991) classifies uncertainty causes into two categories: indeterminacy, which reflects the limitations of available information, and incompleteness, due to a simplified representation that allows only partial use of available information—or in simpler terms, either the information or the model. He then subdivides the causes into 14 causes, five of indeterminacy and nine of model incompleteness (Table 1.1).

The rest of this section catalogs these causes (Walley, 1991; Zimmermann, 2011):

- Lack of information: this is the most common cause of uncertainty. Decision making is the perfect example. Indeed, if a decision maker manages to obtain all the necessary data and consider the different possible outcomes and consequences, he will surely choose the solution he considers the best, but this is rarely the case. One can distinguish the extreme case that would be the absence of information, total ignorance, launching a new project, sending a space shuttle to a planet never visited. The solution would be to have more

TABLE 1.1
Sources of Uncertainty (Limbourg, 2008)

Indeterminacy	Incompleteness
Lack of information	Lack of introspection
Conflicting information	Lack of assessment strategies
Conflicting beliefs	Limits in computational ability
Information of limited relevance	Intractable models
Physical indeterminacy	Natural extension
	Choice of elicitation structure
	Ambiguity
	Instability
	Caution in elicitation

information. However, according to Heisenberg's uncertainty principle "we cannot get all the information we need, since the act of obtaining information often alters the phenomena studied" (Heisenberg, 1971).
- We can also add in opposition to lack of information, too much information, called in (Zimmermann 2011) "Abundance of information (complexity)": in order to fill the gap, we try to have enough information. However "too much information kills information." Spira (2011) explains how too many messages, email and so forth, can be harmful to an organization. Moore's empirical law posits that the situation can only get worse. This law stipulates in its first version that the average number of transistors in a microprocessor doubles every two years. It was then revised to state that a given something—in this case, data—doubles every 18 months. Any researcher can experiment this by typing the title of their study on a search engine; they will find that this prediction is approximately true (Coffman & Odlyzko, 2002), making a literature review, at best, difficult. (Note, though, that Moore's law should be called a conjecture instead, since it is an unproven assertion.) In addition, too much information can change the way we think (Andrejevic, 2013). We have the false impression that we can know everything through a few clicks on the internet. However, can we say with certainty that the information found is the right one! In a historical view of information management, Blair (2010) points out the need to review the way information is managed in general and surplus information in particular.
- Conflicting information: in situations where information conflicts, instead of adding more information—which can worsen the situation, it is better to look for false information that has not been identified as false. The conflict may be due to different states studied that cannot be distinguished.
- Conflicting beliefs: contrary to conflicting information, conflicting beliefs are not based on evidence, but on expert opinion.
- Information of limited relevance: the information on which the model is based is hardly usable for prediction.
- Physical indeterminacy: the amount of interest is not precisely predictable, even with infinite information.
- Lack of introspection: this can come from lack of time or resources or simply because it is not worth the cost, what Walley (1991) calls the "cost of thought."
- Lack of evaluation strategies: relevant information is available, but it is not easy to introduce into the model (for example, text reports). If the data is transformed, the uncertainty increases. We may have a lot of relevant information but be unable to use it because we are missing the necessary models or evaluation strategies. It is often necessary to simplify, idealize, or ignore some of the evidence, or to make gross overall judgments of probabilities. Additional inaccuracy should be introduced to allow these distortions.
- Limits in computational ability: determining an exact value can be difficult, but getting adequate upper and lower limits can be less tricky.
- Intractable models: the actual model may be inconvenient or too complex, but replacing it with a simpler one may only increase uncertainty.
- Natural extension: while facing new problems, human nature tends to favor adapting existing models rather than building new ones. And since the basic

model is only an idealization of a real situation, its extension, at best, will include the uncertainties of the previous model, to which will be added those due to the extension.
- Choice of elicitation structure: depending on the evaluator's experience and evaluation technique, there may be a limit to accuracy.
- Ambiguity: natural language ambiguity is transferable to the scientific field. Quantitative translation of expressions such as "probably" and "around" can introduce uncertainty. According to Zimmermann (2011) this can be seen as a lack of information, because clarifying the speech context allow us to better interpret the meaning of each word. However, the obligatory transition to the quantitative makes the vague aspect emerge. All the more, words change meaning over time, sometimes radically. It has already occurred, and nothing prevents more change.
- Instability: using different methods or even the same method twice can lead to different outcomes.
- Caution in elicitation: or simply "Modeler precaution": wanting to avoid any possible later reproaches, engineers or anyone responsible for modeling will prefer to indicate a greater error interval than necessary. This is also true for new projects, which is consistent with the absence or scarcity of information, and in partnerships between several stakeholders.
- Acknowledging that they belong to the first category (lack of information), Zimmermann adds measures and beliefs. Measures matter because of their importance in the framework of engineering, which justifies holding the quadrennial General Conference on Weights and Measures. Beliefs, on the other hand, derive their legitimacy from practical cases, where all the information presented is subjective.

INFORMATION TYPES

According to Zimmermann (2011), after identifying uncertainty causes we must know the types of information available, which he roughly classifies as numerical (cardinal, ordinal, nominal), linguistic (in natural language), interval, and symbolic. He insists that the treatment of any given piece of information should correspond to its type.

- Numerical: this information can be cardinal, ordinal, nominal, or in the form of a ratio.
- Linguistic: this is any information given in natural language, not in a formal language. In this type of situation, the word, such as a label, must be distinguished from its meaning, because there is no binary relationship between a word and its meaning nor a measure of the quality of such information. Language information has developed as a means of communication between human beings whose "inference engines" are their own minds, which are still little known.
- Interval: to treat interval information correctly, we have to use adequate arithmetic that will lead to an output of the same type. Regardless of how it was acquired, the information obtained is considered exact.

- Symbolic: the information is as valuable as the definitions of the symbols used. Furthermore, the type of information processing must also be symbolic rather than numerical nor linguistic.

UNCERTAINTY THEORIES

The result of tossing a coin or rolling a die and other situations can be perfectly modeled by probabilities, and this book does not seek to impose otherwise. In other words, this part is neither a plea for fuzzy Logic nor an indictment against probability. Moreover, one should try to avoid reductive visions such as probabilistic, non-probabilistic (Wu, Apostolakis, & Okrent, 1990), or a probabilistic/possibilistic duality (Klir, 1990). The distinction between probability and possibility can be summed up as follows: "possibility can be interpreted as a response to the question: can it happen? Probability is an answer to the question: how often?" (Raskin & Taylor, 2014).

Until the 1960s, there was some equivalence between probability and uncertainty. This historical dominance is palpable in textbooks; the reader is likely more familiar with probability than any other theory dealing with uncertainty. Since then, it has been found that probability alone cannot address all forms of uncertainty (Klir, 1989). This has given rise to several theories—including fuzzy set theory (Zadeh, 1965), theory of evidence (Dempster, 1967a, 1967b, 1968; Shafer, 1976), possibility theory (Zadeh, 1978), theory of fuzzy measurements (Sugeno, 1993, 1974), theory of rough sets, (Pawlak, 1982), and intuitionistic fuzzy sets (Atanassov, 1986, 1999; Atanassov & Gargov, 1989). These recent theories stimulants an increasing number of researchers around the world.

The obvious question is: Why so many theories? It should be noted first that some of these theories generalize others (i.e. not all of them are mutually exclusive). However, none of them manages to model uncertainty in all its aspects (Zimmermann, 2011). This justifies the existence of attempts to develop generalizing theories that encompass uncertainty in all its aspects (Baldwin, 1991; Klir & Wierman, 2013; Zadeh, 2006).

Without pretending to be exhaustive, Zimmermann stipulates that the chosen theory must take into consideration four criteria: uncertainty cause, type of information available (input information), numerical information type, and type of information desired (output information) (Zimmermann, 2011). He adds that the combination adapted to a probabilistic presentation brings together uncertainty resulting from a lack of information and numerical information of the cardinal type in and out of the system studied. Other combinations would be better suited to other theories of uncertainty (Figure 1.1).

Klir (2005) points out that in order to develop a fully operational theory to deal with the uncertainty of a certain type, a multitude of issues must be addressed at each of the following four levels:

- Level 1: an appropriate mathematical formalization of the designed type of uncertainty must be found.
- Level 2: a calculation must be developed by which this type of uncertainty can be handled correctly.

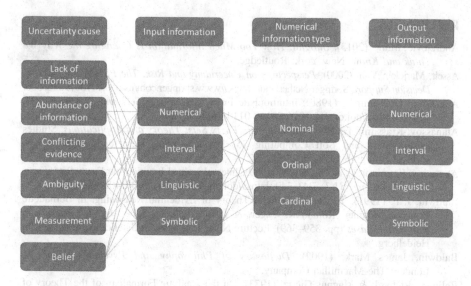

FIGURE 1.1 Rough Classification of Properties of Uncertainty (Zimmermann, 2011).

- Level 3: a meaningful way must be found to measure the amount of uncertainty relevant to any situation that can be formalized in the theory.
- Level 4: methodological aspects of the theory must be developed, including procedures to make the various principles of uncertainty operational within the theory.

Note that in general, uncertainty has already undergone a transformation before being presented to the user. Hence, Klir speaks of two uncertainties: before and after transformation.

CONCLUSION

This chapter emphasizes the essentiality of uncertainty while differentiating ambiguity and vagueness in the logical sense, which fuzzy logic allows to deal with simultaneously. The chapter maintains that instead of looking for an exact description of the world, it is more reasonable and realistic, or even desirable in some situations, to reduce the level of uncertainty instead. Then it focuses on distinguishing the 14 causes of uncertainty, five of which generate data indeterminacy and the remaining nine generate model incompleteness; in addition to three additional causes. After that, it discusses the types of information manipulated—numerical, interval, linguistic, and symbolic—and the models for processing them. Subsequently, based on Zimmermann's work, it stresses that each situation must be linked to the most appropriate theory of uncertainty, since no theory can capture all aspects. The chapter concludes by raising the question of the necessary precepts to build a theory of uncertainty, while insisting that the obtained result is only a broad representation of reality.

REFERENCES

Andrejevic, Mark. (2013). *Infoglut: How Too Much Information Is Changing the Way We Think and Know*. New York: Routledge.
Asselt, Marjolein Van. (2000). *Perspectives on Uncertainty and Risk: The PRIMA Approach to Decision Support*. Springer Netherlands. https://www.springer.com/us/book/9780792366560.
Atanassov, Krassimir T. (1986). Intuitionistic Fuzzy Sets. *Fuzzy Sets and Systems*, *20*(1), 87–96. https://doi.org/10.1016/S0165-0114(86)80034-3.
Atanassov, Krassimir T. (1999). *Intuitionistic Fuzzy Sets: Theory and Applications*. Studies in Fuzziness and Soft Computing 35. Physica-Verlag Heidelberg. https://www.springer.com/gp/book/9783790812282.
Atanassov, Krassimir T., & G. Gargov. (1989). Interval Valued Intuitionistic Fuzzy Sets. *Fuzzy Sets and Systems*, *31*(3), 343–349. https://doi.org/10.1016/0165-0114(89)90205-4.
Baldwin, J. F. (1991). Towards a General Theory of Evidential Reasoning. In Bernadette Bouchon-Meunier, Ronald R. Yager, & Lotfi A. Zadeh edited by, *Uncertainty in Knowledge Bases* (pp. 359–369). Lecture Notes in Computer Science. Springer Berlin Heidelberg.
Baldwin, James Mark. (1902). *Dictionary of Philosophy and Psychology, Vol. II*. London: The Macmillan Company.
Bellman, Richard, & Magnus Giertz. (1973). On the Analytic Formalism of the Theory of Fuzzy Sets. *Information sciences*, *5* (January), 149–156. https://doi.org/10.1016/0020-0255(73)90009-1.
Berlinghoff, William P., Kerry E. Grant, & Dale Skrien. (2001). *A Mathematics Sampler: Topics for the Liberal Arts*. USA: Rowman & Littlefield.
Black, Max. (1963). Reasoning with Loose Concepts. *Dialogue: Canadian Philosophical Review / Revue Canadienne de Philosophie*, *2*(1), 1–12. https://doi.org/10.1017/S001221730004083X.
Blair, Ann M. (2010). *Too Much to Know: Managing Scholarly Information before the Modern Age*. London: Yale University Press.
Borislav, Demitrov. (2018). *Topological (in) Hegel: Topological Notions of Qualitative Quantity and Multiplicity in Hegel's Fourfold of Infinities*. Sofia, Bulgaria: Borislav Dimitrov.
Braun, Matthias, Hannah Schickl, & Peter Dabrock. (Eds.), (2018). *Between Moral Hazard and Legal Uncertainty: Ethical, Legal and Societal Challenges of Human Genome Editing*. Technikzukünfte, Wissenschaft Und Gesellschaft / Futures of Technology, Science and Society. VS Verlag für Sozialwissenschaften. https://www.springer.com/us/book/9783658226596.
Coffman, K. G., & A. M. Odlyzko. (2002). Internet Growth: Is There a 'Moore's Law' for Data Traffic? In James Abello, Panos M. Pardalos, & Mauricio G. C. Resende edited by, *Handbook of Massive Data Sets* (pp. 47–93). Massive Computing. Boston, MA: Springer US. https://doi.org/10.1007/978-1-4615-0005-6_3.
Deemter, Kees van. (2010). *Not Exactly: In Praise of Vagueness*. Great Britain: OUP Oxford.
Dempster, A. P. (1967a). Upper and Lower Probabilities Induced by a Multivalued Mapping. *The Annals of Mathematical Statistics*, *38*(2), 325–339.
Dempster, A. P. (1967b). Upper and Lower Probability Inferences Based on a Sample from a Finite Univariate Population. *Biometrika*, *54*(3–4), 515–528. https://doi.org/10.1093/biomet/54.3-4.515.
Dempster, A. P. (1968). A Generalization of Bayesian Inference. *Journal of the Royal Statistical Society. Series B (Methodological)*, *30*(2), 205–247.
Dompere, Kofi Kissi. (2013). *Fuzziness and Foundations of Exact and Inexact Sciences*. Studies in Fuzziness and Soft Computing. Berlin, Heidelberg: Springer-Verlag. https://www.springer.com/la/book/9783642311215.

Ebert, Philip A., & Marcus Rossberg. (2016). *Gottlob Frege: Basic Laws of Arithmetic*. Reprint. Oxford: OUP Oxford.

Grant, Edward. (2007). *A History of Natural Philosophy: From the Ancient World to the Nineteenth Century*. 1 edition. New York: Cambridge University Press.

Heisenberg, Werner. (1971). *Physique et philosophie: la science moderne en révolution*. France: Albin Michel.

Helton, Jon C. (1994). Treatment of Uncertainty in Performance Assessments for Complex Systems. *Risk Analysis*, *14*(4), 483–511. https://doi.org/10.1111/j.1539-6924.1994.tb00266.x.

Henrion, Max, & Baruch Fischhoff. (1986). Assessing Uncertainty in Physical Constants. *American Journal of Physics*, *54*(9), 791–798. https://doi.org/10.1119/1.14447.

Hoyt, John P. (1941). Examples of Inexactness in an Exact Science. *School Science and Mathematics*, *41*(7), 627–628. https://doi.org/10.1111/j.1949-8594.1941.tb05463.x.

Inuiguchi, Masahiro, & Jaroslav Ramík. (2000). Possibilistic Linear Programming: A Brief Review of Fuzzy Mathematical Programming and a Comparison with Stochastic Programming in Portfolio Selection Problem. *Fuzzy Sets and Systems*, *111*(1), 3–28. https://doi.org/10.1016/S0165-0114(98)00449-7.

Jourdain, Philip E. B. (1912). Mr. Bertrand Russell's First Work on the Principles of Mathematics. *The Monist*, *22*(1), 149–158. https://doi.org/10.5840/monist19122213.

Kahneman, Daniel, & Amos Tversky. (1982). Variants of Uncertainty. *Cognition*, *11*(2), 143–157. https://doi.org/10.1016/0010-0277(82)90023-3.

Kattsoff, Louis O. (1957). Presuppositions of Science. In Louis O. Kattsoff edited by, *Physical Science and Physical Reality* (pp. 274–289). Dordrecht: Springer Netherlands. https://doi.org/10.1007/978-94-017-6048-5_19.

Kaufmann, A. (1975). *Introduction to the Theory of Fuzzy Subsets*. 1st edition. New York: Academic Press.

Keil, Geert, & Ralf Poscher. (2016). *Vagueness and Law: Philosophical and Legal Perspectives*. United Kingdom: Oxford University Press.

Klir, G. J. (1987). Where Do We Stand on Measures of Uncertainty, Ambiguity, Fuzziness, and the Like? *Fuzzy Sets and Systems*, Measures of Uncertainty, *24*(2), 141–160. https://doi.org/10.1016/0165-0114(87)90087-X.

Klir, G. J. (1989). Is There More to Uncertainty Than Some Probability Theorists Might Have Us Believe? *International Journal of General Systems*, *15*(4), 347–378. https://doi.org/10.1080/03081078908935057.

Klir, G. J. (1990). Probabilistic versus Possibilistic Conceptualization of Uncertainty. In *[1990] Proceedings. First International Symposium on Uncertainty Modeling and Analysis*, 38–41. https://doi.org/10.1109/ISUMA.1990.151219.

Klir, George J. (2005). *Uncertainty and Information: Foundations of Generalized Information Theory*. USA: John Wiley & Sons.

Klir, George J., & Mark J. Wierman. (2013). *Uncertainty-Based Information: Elements of Generalized Information Theory*. Berlin: Physica.

Köhn, Julia. (2017). *Uncertainty in Economics: A New Approach*. Switzerland: Springer.

Koopmans, Tjalling Charles. (1957). *Three Essays on the State of Economic Science*. London: McGraw-Hill.

Li, Y., J. Chen, & L. Feng. (2013). Dealing with Uncertainty: A Survey of Theories and Practices. *IEEE Transactions on Knowledge and Data Engineering*, *25*(11), 2463–2482. https://doi.org/10.1109/TKDE.2012.179.

Limbourg, Philipp. (2008). *Dependability Modelling under Uncertainty: An Imprecise Probabilistic Approach*. Studies in Computational Intelligence. Berlin, Heidelberg: Springer-Verlag. https://www.springer.com/gp/book/9783540692867.

Natke, H. Günther, & Yakov Ben-Haim. (1997). *Uncertainty: Models and Measures*: Proceedings of the International Workshop Held in Lambrecht, Germany, July 22–24, 1996. Australia: Akademie Verlag.
Nemchinov, Academician V. (1962). Economic Science Must Become an Exact Science. *Øst-Økonomi*, *2*(1), 36–44. https://doi.org/10.1007/BF02506033.
Neugebauer, Otto. (1969). *The Exact Sciences in Antiquity*. USA: Courier Corporation.
Nicolai, Alexander T., & Jörg M. Dautwiz. (2010). Fuzziness in Action: What Consequences Has the Linguistic Ambiguity of the Core Competence Concept for Organizational Usage? *British Journal of Management*, *21*(4), 874–888. https://doi.org/10.1111/j.1467-8551.2009.00662.x.
Pawlak, Zdzisław. (1982). Rough Sets. *International Journal of Computer & Information Sciences*, *11*(5), 341–356. https://doi.org/10.1007/BF01001956.
Planck, Max. (1949). The Meaning and Limits of Exact Science. *Science*, *110*(2857), 319–327. https://doi.org/10.1126/science.110.2857.319.
Raskin, V., & J. M. Taylor. (2014). Fuzziness, Uncertainty, Vagueness, Possibility, and Probability in Natural Language. In *2014 IEEE Conference on Norbert Wiener in the 21st Century (21CW)*, USA, 1–6. https://doi.org/10.1109/NORBERT.2014.6893868.
Reduction, National Research Council (U S.) Committee on Risk-Based Analysis for Flood Damage. (2000). *Risk Analysis and Uncertainty in Flood Damage Reduction Studies*. National Academy Press.
Ronen, Yigal. (1988). *Uncertainty Analysis*. 1 edition. Boca Raton, FL: CRC Press.
Russell, Bertrand. (1923). Vagueness. *Australasian Journal of Philosophy*, *1*(2), 84–92.
Russell, Bertrand. (1951). *The Wit and Wisdom of Bertrand Russell*. USA: Beacon Press.
Sallak, Mohamed, Felipe Aguirre, & Walter Schon. (2013). Incertitudes Aléatoires et Épistémiques, Comment Les Distinguer et Les Manipuler Dans Les Études de Fiabilité? In *QUALITA2013*. Compiègne, France. https://hal.archives-ouvertes.fr/hal-00823114.
Sanford, David H. (1995). Uses and Abuses of Fuzziness in Philosophy: A Selective Survey of How Recent Philosophical Logic and Metaphysics Treat Vagueness. *International Journal of General Systems*, *23*(3), 271–277. https://doi.org/10.1080/03081079508908043.
Schomberg, René von. (1993). Controversies and Political Decision Making. In René Von Schomberg edited by, *Science, Politics and Morality: Scientific Uncertainty and Decision Making* (pp. 7–26). Theory and Decision Library. Dordrecht: Springer Netherlands. https://doi.org/10.1007/978-94-015-8143-1_2.
Shafer, Glenn. (1976). *A Mathematical Theory of Evidence*. USA: Princeton University Press.
Smart, J. J. C. (1959). Can Biology Be an Exact Science? *Synthese*, *11*(4), 359–368.
Spira, Jonathan B. (2011). *Overload!: How Too Much Information Is Hazardous to Your Organization*. John Wiley & Sons.
Sugeno, M. (1974). *Theory of Fuzzy Integrals and Its Applications*. Tokyo: Tokyo Institute of Technology.
Sugeno, M. (1993). Fuzzy Measures and Fuzzy Integrals—A Survey. In Didier Dubois, Henri Prade, & Ronald R. Yager edited by, *Readings in Fuzzy Sets for Intelligent Systems* (pp. 251–57). Morgan Kaufmann. https://doi.org/10.1016/B978-1-4832-1450-4.50027-4.
Sutton, O. G. (1954). The Development of Meteorology as an Exact Science. *Quarterly Journal of the Royal Meteorological Society*, *80*(345), 328–338. https://doi.org/10.1002/qj.49708034503.
Tchougréeff, Andrei L. (2016). Several Stories from Theoretical Chemistry with Some Russian Flavor and Implications for Theorems of Chemistry, Vagueness of Its Concepts, Fuzziness of Its Definitions, Iconicity of Its Language, and Peculiarities of

Its Nomenclature. *International Journal of Quantum Chemistry*, *116*(3), 137–160. https://doi.org/10.1002/qua.25050.

Venrooij, Alex van, & Vaughn Schmutz. (2018). Categorical Ambiguity in Cultural Fields: The Effects of Genre Fuzziness in Popular Music. *Poetics*, *66*(February), 1–18. https://doi.org/10.1016/j.poetic.2018.02.001.

Walley, Peter. (1991). *Statistical Reasoning with Imprecise Probabilities*. Australia: Taylor & Francis.

Watson, E. E. (1939). Progress in Exact Science. *School Science and Mathematics*, *39*(9), 824–836. https://doi.org/10.1111/j.1949-8594.1939.tb04032.x.

Williamson, Timothy. (1994). *Vagueness*. New York, NY: Routledge.

Wu, J. S., G. E. Apostolakis, & D. Okrent. (1990). Uncertainties in System Analysis: Probabilistic versus Nonprobabilistic Theories. *Reliability Engineering & System Safety*, *30*(1), 163–181. https://doi.org/10.1016/0951-8320(90)90093-3.

Zadeh, L. A. (1965). Fuzzy Sets. *Information and Control*, *8*(3), 338–353. https://doi.org/10.1016/S0019-9958(65)90241-X.

Zadeh, L. A. (1978). Fuzzy Sets as a Basis for a Theory of Possibility. *Fuzzy Sets and Systems*, *1*(1), 3–28. https://doi.org/10.1016/0165-0114(78)90029-5.

Zadeh, L. A. (2006). Generalized Theory of Uncertainty (GTU): Principal Concepts and Ideas. *Computational statistics & data analysis*, The Fuzzy Approach to Statistical Analysis, *51*(1), 15–46. https://doi.org/10.1016/j.csda.2006.04.029.

Zhang, Qiao. (1998). Fuzziness - Vagueness - Generality - Ambiguity. *Journal of Pragmatics*, *29*(1), 13–31. https://doi.org/10.1016/S0378-2166(97)00014-3.

Zimmermann, Hans-Jürgen. (2011). *Fuzzy Set Theory—and Its Applications*. New York: Springer Science & Business Media.

2 Nonclassical Logics

LIMITS OF CLASSICAL LOGIC

Experience shows that out of a thousand young people who learn logic, not ten know anything about it six months after they have completed their courses (Arnauld & Nicole, 1992)!

Theoretically, a logic contains three aspects:

- A symbolic language allowing the formal representation of statements
- A formal deduction system
- A link between formulas and certain realities (semantics)

Logic is most often related to the field of mathematics, made by and for mathematicians; all the more, all the reasoning chosen and the choices made acquire their legitimacy from the logic followed. The most often recurring and taught logic is that formalized by Boole (2016).

However, Boolean logic, which recognizes only two states of nature—the absolutely true "1" and the absolutely false "0," rejects all kinds of transient states. Does this system of reasoning represent nature as it is? A speech attributed to Albert Einstein states that "the more categorical mathematics is, the further away it is from reality."

Take the case of a young adult aged 17 years and 364 days (Figure 2.1). According to the Boolean logic, this young person is considered a minor, but the following day this same person is considered major (an adult). This same distinction is observed in law, in terms of the consequences that can result from civil offenses (civil liability), access to work, and other domains. Does that seem consistent? Is personality forged in 24 hours or less? Does such a short time allow such a significant change?

The most popular example of a logic that admits several states—conflicting in the Boolean sense—is Schrödinger's cat (Gribbin, 2008). This cat is considered both dead and alive, and several mathematical theories can affirm the merits of this theory; but how can it exist? According to common sense, such situation would be purely philosophical.

Moreover, logic was born in the vicinity of philosophy, and this link remained until the middle of the 19th century. Nevertheless, the reformers of traditional logic who gave birth to mathematical logic were mathematicians, who developed it from deep philosophical reflection. However, it is difficult to give a definition of logic. In fact, is there any meaning in talking about logic? Wouldn't there be logics instead? If so, what keeps them separate and what brings them together (Belna, 2014)? A logic is concerned with the study of reasoning, its modeling, and its validation.

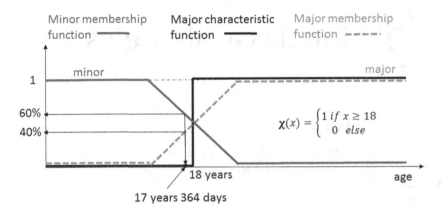

FIGURE 2.1 Minor/ Major Discernibility in Classical and Fuzzy Logic.

We can only speak of "logic" in the particular case of one of the vast areas of philosophy; otherwise, we must speak of "logics."

The remainder of this chapter will be about the limits of the binary parity true/false of classical logic. It will also be about some unconventional logics admitting several truth values, as well as the reasons hindering their understanding and use.

MULTIVALUED LOGICS

Usually, Aristotelian logic is confined to the Boolean syllogism (Gourinat, 2011). However, multivalued logics—called also many- or multiple-valued logics—go a long way back in time, as early as the fourth century BC. Aristotle, in the fourth chapter of *De Interpretatione*, enunciates "there will be a naval battle tomorrow" in binary logic; his negation is "there will not be a naval battle tomorrow," and therefore one or the other is necessarily true. However, can it be concluded that the future event itself is necessary (Jolivet, 1994)? In advancing such formulations, Aristotle wanted to demonstrate the absurd consequences of bivalence (Husson, 2009), suggesting the existence of a third truth value.

TRIVALENT LOGICS

ŁUKASIEWICZ'S TRIVALENT LOGIC

The revival of multivalued logics began in the 1920s, notably through the work of Łukasiewicz (1920) and Post (1921). Similar to Aristotle, the Polish philosopher Łukasiewicz asserted that the statement "I will be in Warsaw at noon on December 21 next year" is neither true nor false at the time of the enunciation, and that it represents only a possibility, not a necessity. If the statement were true at the present moment, the speaker's future presence in Warsaw is necessary, contrary to Łukasiewicz's assertion. In the opposite case (if the assertion were false at the time of enunciation), the future presence of the speaker would be impossible. Hence, neither case is certain, even if they contradict each other (Łukasiewicz, 2000).

Nonclassical Logics

Łukasiewicz (1920) denoted this third value 1/2 (after discarding his initial notation of 2), keeping the classic values of 1 for true and 0 for false. This third value has been interpreted as possibility or indeterminacy. This extends negation ($\neg$), implication ($\rightarrow$), disjunction ($\vee$), conjunction ($\wedge$), and equivalence ($\equiv$) as in Tables 2.1–2.5.

It should be noted that Łukasiewicz's logic requires that the identity law $P \rightarrow P$ be always verified.

Kleene's Trivalent Logic

In addition to the classical true and false, Kleene (1938) introduces a third value which he calls "indefinite." This value is attributed to a proposition P when neither P nor $\neg P$ can be demonstrated. In order to facilitate comparison, true, indefinite, and false are denoted as before: respectively 1, 1/2, and 0.

TABLE 2.1
Łukasiewicz's Negation

P	$\neg P$
0	1
1/2	1/2
1	0

TABLE 2.2
Łukasiewicz's Implication

$\rightarrow$	0	1/2	1
0	1	1	1
1/2	1/2	1	1
1	0	1/2	1

TABLE 2.3
Łukasiewicz's Disjunction

$\vee$	0	1/2	1
0	0	1/2	1
1/2	1/2	1/2	1
1	1	1	1

TABLE 2.4
Łukasiewicz's Conjunction

∧	0	1/2	1
0	0	0	0
1/2	0	1/2	1/2
1	0	1/2	1

TABLE 2.5
Łukasiewicz's Equivalence

≡	0	1/2	1
0	1	1/2	0
1/2	1/2	1	1/2
1	0	1/2	1

While Kleene's negation is identical to that proposed by Łukasiewicz, Kleene defines implication as a generalization of classical implication $P \to Q \equiv \neg P \vee Q$. In Kleene's vision, as soon as indefiniteness intervenes in the calculation, the result is automatically indefinite too. The truth tables remain similar Łukasiewicz's, except for boxes involving an indeterminate value (Tables 2.6–2.9).

Kleene (1952) later introduced a second trivalent logic that he designated "weak," and called the previous one by "strong." Fitting and Orlowska (2003) explain that the difference between the two is illustrated by the difference between the conjunctions & and && and the disjunctions | and || used in programming. The truth tables of Kleene's strong logic are identical to Łukasiewicz's, except for implication (Table 2.10).

Bochvar's Trivalent Logic

Bochvar was inspired by paradoxes (Szmielew, 1946), such as the liar paradox (Bergmann, 2008). If a person tells you that he is lying to you, then reasoning

TABLE 2.6
Kleene's Strong Implication

→	0	1/2	1
0	1	1	1
1/2	1/2	1/2	1
1	0	1/2	1

TABLE 2.7
Kleene's Disjunction (∨)

∨	0	1/2	1
0	0	1/2	1
1/2	1/2	1/2	1/2
1	1	1/2	1

TABLE 2.8
Kleene's Conjunction

∧	0	1/2	1
0	0	1/2	0
1/2	1/2	1/2	1/2
1	0	1/2	1

TABLE 2.9
Kleene's Equivalence

≡	0	1/2	1
0	1	1/2	0
1/2	1/2	1/2	1/2
1	0	1/2	1

TABLE 2.10
Kleene's Implication

→	0	1/2	1
0	1	1/2	1
1/2	1/2	1/2	1/2
1	0	1/2	1

classically, with two possible values, either what the person has just said is true and therefore the person is lying or the sentence is false and therefore the person is telling the truth. According to Bochvar, such a situation is insignificant, "meaningless" (Church, 1939). He distinguishes the statement from the sentence: if a statement is logical, it is necessarily true or false, while a sentence may be true, false, or insignificant (Bergmann, 2008).

While the truth tables of the Bochvar's internal logic are identical to those of Kleene's logic, those of Bochvar's external logic (Bochvar, 1984) are clearly different from all of the preceding ones (Tables 2.11–2.15).

TABLE 2.11
Bochvar's Negation

P	¬P
0	1
1/2	1
1	0

TABLE 2.12
Bochvar's Implication

→	0	1/2	1
0	1	1	1
1/2	1	1	1
1	0	0	1

TABLE 2.13
Bochvar's Disjunction

∨	0	1/2	1
0	0	0	1
1/2	0	0	1
1	1	1	1

TABLE 2.14
Bochvar's Conjunction

∧	0	1/2	1
0	0	0	0
1/2	0	0	0
1	0	0	1

TABLE 2.15
Bochvar's Equivalence

≡	0	1/2	1
0	1	1	0
1/2	1	1	0
1	0	0	1

Other nonsense logics have been presented (Âqvist, 1962; Hallden, 1949; Piróg-Rzepecka, 1977; Reichenbach, 1946; Segerberg, 1965). A survey focusing on the algebraic aspect was presented by Bolc and Borowik (1992). A simple question can be asked: Why so many trivalent logics? The answer lies in their various applications. Kleene's logic (Kleene, 1952) was seen as a tool to address several issues concerning recursive partial functions. Rasiowa (1974) investigated the applicability of trivalent logic in surveys of the accuracy of computational programs. Piróg-Rzepecka's approach (Piróg-Rzepecka, 1977) provides a method for analyzing and using a trivalent logic in some mathematical problems involving expressions that may lack meaning. Questions of a purely logical nature have been dealt with by Baylis (1937, 1938) and Tarski (1937). Other questions dealing with quantum physics are presented by Birkhoff and Von Neumann (1936) and Destouches-Février (1948).

N-VALUED LOGICS (N > 2)

In our quest for a better representation, is a trivalent logic sufficient? Even if the philosophical side remains pervasive, an attachment to reality is always welcome. Moreover, since we live in a consumer society, the answer to how many values are required can be: How many do you want? Alternatively, how many do you need? (Fitting & Orlowska, 2003) It is in this spirit that Łukasiewicz generalized his trivalent logic to a multivalent one (Lukasiewicz, 1930; Lukasiewicz, Borkowski, & Wojtasiewicz, 1970) that can take more than three possible values verifying

$$\frac{k}{n-1}, \quad 0 \leq k \leq n-1.$$

While truth tables can easily be drawn for relatively small n, the task is less evident as n gets bigger. Fortunately, Łukasiewicz's algebraic expressions allow for a condensed representation. Denote by $\tau(P)$ and $\tau(Q)$ the truth values of P and Q, respectively; then

$$\tau(\neg P) = 1 - \tau(P),$$

$$\tau(P \to Q) = \min\{1 - \tau(P) + \tau(Q), 1\},$$

$$\tau(P \vee Q) = \max\{\tau(P), \tau(Q)\},$$

$$\tau(P \wedge Q) = \min\{\tau(P), \tau(Q)\},$$

$$\tau(P \equiv Q) = 1 - |\tau(P) - \tau(Q)|.$$

It is easy to verify that these equations achieve the same results as in Tables 2.1–2.5 for $n = 3$.

Alongside Łukasiewicz's work, Post (1921) focuses on the formal algebraic aspect, without worrying too much about the philosophical side or the interpretation of nonclassical truth values. He defines for any natural number greater than or equal to 2 a set of linearly ordered truth values $\{t_1, \ldots, t_n\}$ whose limit values t_1 and t_n represent the absolutely true and the absolutely false. Drawing on the work of Whitehead and Russell (1910), he bases his approach on the negation $\neg$ and disjunction $\vee$, which he defines as follows:

$$\neg t_i = \begin{cases} t_{i+1} \text{si} & i \neq n \\ t_1 \text{si} & i = n \end{cases}$$

$$t_i \vee t_j = t_{\max\{i,j\}}.$$

for $n = 4$, the negation (Table 2.16) and the disjunction (Table 2.17) are as follows:

TABLE 2.16
Multivalent Negation

P	¬P
t_1	t_2
t_2	t_3
t_3	t_4
t_4	t_1

TABLE 2.17
Multivalent Disjunction

$\vee$	t_1	t_2	t_3	t_4
t_1	t_1	t_2	t_3	t_4
t_2	t_2	t_2	t_3	t_4
t_3	t_3	t_3	t_3	t_4
t_4	t_4	t_4	t_4	t_4

Applications of Post's logic have been proposed in computer science (Epstein, Frieder, & Rine, 1974; Rine, 2014). Other applications of the panoply of multi-valued logics have been presented by Azevedo (2003), Miller and Thornton (2008), Vranesic and Smith (1974), Stanković, Astola, and Moraga, (2012) and Epstein (1993).

The question may still be asked about the optimal number of possible truth values and its relevance. The fuzzy logic that will be developed in the next chapter constitutes the transition from a logic that can only take predetermined truth values in the interval [0, 1], such as those of Łukasiewicz and Post, to a logic that can take any value in the same range (i.e. $n \to +\infty$).

It should be noted that logics with infinite truth values are not only linked to fuzzy logic, since studies concerning them were already present long before the appearance of fuzzy logic in 1965 (Belluce, 1964; Belluce & Chang, 1963; McNaughton, 1951; Turquette, 1963). It was not until 1969 that Goguen established a formal link between the two (Goguen, 1969), and investigations have continued to recent years (Dubois & Toffano, 2017; Gerasimov, 2018; Moussa & Hadj Kacem, 2017).

OTHER NONCLASSICAL LOGICS: QUANTUM MECHANICS

We tend to believe the validity of quantum mechanics, since it explains a reality. However, as soon as you mention that this same framework stipulates that an electron travels two different routes at the same time, that belief is immediately shaken.

In middle school, we were taught the wave/particle duality as established by "the most beautiful experiment in physics" (Crease, 2002), whose roots can be traced back to the scientific quarrel in the 17th century, between Isaac Newton and Christiaan Huygens, over the composition of light. While the former was convinced that light is composed of corpuscles and therefore of particles (the corpuscular theory of light; Newton, 1704), the second affirmed it to be a wave (Huygens, 1678).

In 1801 (Young, 1802), Young's slit experiment demonstrated the undulatory behavior of light. His work, as well as that of Fresnel (1868) and Foucault (1853), helped establish the undulatory theory of light in the 19th century.

Einstein, although aware that his theory (Einstein, 1905) did not explain the interference phenomenon, assures readers that it is the only way to understand the photoelectric effect discovered by Hertz (1887), in which light striking a surface made of certain metals creates electron movement.

When Young's experiment is undertaken with a flow of light allowing the emission of one electron at a time, each of them leaves an individual impact, reminiscent of the corpuscular theory. However, at some point we realize that all the spots constitute interference. The impact is corpuscular, but the behavior is undulatory. Even more spectacular, when trying to ascertain the slot through which each electron passes, the interference goes away. A standard interpretation states that the electron passes through the two slots at the same time. Heisenberg (1962) explains that it is not nature itself being described this way, but nature subject to our method of questioning.

In seeking applications of Łukasiewicz's trivalent logic, Zawirski (1932) asserts that "light is a wave and light is made of particles" can only be envisaged in the presence of a logic permitting at least three truth values. More detailed studies of the different interpretations were developed subsequently (Evans & Thorndike, 2007; Heelan, 2015; Jammer, 1974). The best-known example is Schrödinger's cat. This is a thought experiment in which a cat is placed in a box that also contains a radioactive substance and a poison. The substance has a 1/2 probability of decaying during the experiment, causing the release of the poison and thus the death of the cat. However, if the particle does not decay—probability also 1/2—the cat lives.

According to quantum mechanics, until the box is opened, the particle both has and has not decayed. Therefore, the poison both has and has not been released. Ultimately, the cat is dead and alive at the same time (Trimmer, 1980). On a macroscopic state, this is counterintuitive: by placing a camera inside, one can observe the state of the cat, dead or alive. But in quantum physics an initial explanation was provided by Hugh Everett III—that the entire universe is quantum. Subsequently, superposition of states (Byrne, 2012; Everett, 2012). This was demonstrated by Serge Haroche's team (Raimond & Brune, 2013). What good is this question of a cat that is dead and alive at the same time? It may be used to build quantum computers, which would make today's computers obsolete. (It should be noted that calculability is not questioned.)

Is this overlapping of states a correct representation of reality? Many people doubt it, suggesting that we are just using words we know to explain concepts we have not mastered. At the end of his courses on quantum mechanics, Einstein told his students, "If you understand me, it's because I haven't been clear."

Other unconventional logics have been developed: modal (Gonzalez, 1992), temporal (Mamache, 2012), intuitionistic (Matsumakia, 2010), and many others (Bernadet, 2011; Costa, 1997). A comprehensive study of the evolution of logic is presented in the Handbook of the History of Logic series (Gabbay & Woods, 2004a, 2004b, 2006, 2008a, 2008b, 2009, 2014; Gabbay, Kanamori, & Woods, 2012; Gabbay, Pelletier, & Woods, 2012; Gabbay, Siekmann, & Woods, 2014; Gabbay, Woods, & Hartmann, 2011).

REASONS FOR REJECTION

Despite the undeniable contributions of nonclassical logics in various fields (Pykacz, 2015; Sokolov & Zhdanov, 2019), they are still not as well accepted as they should be. Putnam (1957) mentions the difficulty of teaching trivalent logics to people who think using only two values, forcing the teacher to use a language with two values also, similar to the use of French to teach Francophones English.

The rejection of trivalent logic can also be due to religious perceptions (Rescher, 1968), because it involves theological difficulties: How can there be divine prescience if future statements are neither true nor false? Rescher adds that divine prescience aside, it is difficult to grant different truth values to "it will rain tomorrow" stated on April 12 and "it rained yesterday" stated on April 14.

Ultimately, the nature of the problems requiring the use of these logics complicates their acceptance. Although simplifying quantum mechanics, it remains counterintuitive even for practitioners. Hence the double challenge: understanding the subject being dealt with as well as the appropriate mode of reasoning.

CONCLUSION

This chapter points the limits of the 0/1 duality of classical logic, illustrated through the difficulty of discerning major versus minor and understanding Schrödinger's cat. It maintains that logic has often evolved in the vicinity of philosophy and mathematics. This rapprochement has made it possible to overcome the Boolean syllogism through successive extensions into multivalent logics. While the first work on this subject dates back to Aristotle, multivalent logics like those of Łukasiewicz, Kleene, and Bochvar—also called multivalued or many-valued logics—are based on the work of Łukasiewicz and Post in the 1920s. Their differences are mainly in their interpretations of negation, implication, disjunction, conjunction, and equivalence.

Driven by a philosophical impulse, Łukasiewicz extended trivalent logic into four-valued and then n-valued logics. Post, for his part, focused on the formal mathematical aspect of these extensions, distancing himself from the philosophical aspect. While Łukasiewicz's logics can only take n predefined values, fuzzy logic can take any value in the interval [0, 1].

The chapter also contains a brief overview of quantum mechanics as a nonclassical logic, through wave/particle duality and Schrödinger's cat, then cites other nonclassical logics and mentions some reasons for their rejection.

REFERENCES

Âqvist, Lennart. (1962). Reflections on the Logic of Nonsense. *Theoria*, 28(2), 138–157.
Arnauld, Antoine, & Pierre Nicole. (1992). *La Logique ou L'art de penser*. Paris: Gallimard.
Azevedo, Francisco. (2003). *Constraint Solving Over Multi-Valued Logics: Application to Digital Circuits*. Netherlands: IOS Press.
Baylis, Charles A. (1937). Jerzy Słupecki. Der Volle Dreiwertige Aussagenkalkül. Comptes Rendus Des Séances de La Société Des Sciences et Des Lettres de Varsovie, Classe III, Vol. 29 (1936), Pp. 9–11. *The Journal of Symbolic Logic*, 2(1), 46 –46. http://dx.doi.org/10.1017/S0022481200039682.
Baylis, Charles A. (1938). Webb Donald L. The Algebra of N-Valued Logic. Comptes Rendus Des Séances de La Société Des Sciences El Des Lettres de Varsovie, Classe III, Vol. 29, Pp. 153–168. *Journal of Symbolic Logic*, 3(1), 52.
Belluce, L. P. (1964). Further Results on Infinite Valued Predicate Logic. *The Journal of Symbolic Logic*, 29(2), 69–78. http://dx.doi.org/10.2307/2270410.
Belluce, L. P., & C. C. Chang. (1963). A Weak Completeness Theorem for Infinite Valued First-Order Logic. *The Journal of Symbolic Logic*, 28(1), 43–50. http://dx.doi.org/10.2307/2271335.

Belna, Jean-Pierre. (2014). *Histoire de la logique*. France: Ellipses.

Bergmann, Merrie. (2008). *An Introduction to Many-Valued and Fuzzy Logic: Semantics, Algebras, and Derivation Systems*. United Kingdom: Cambridge University Press.

Bernadet, Maurice. (2011). *Introduction pratique aux logiques non classiques*. France: Editions Hermann.

Birkhoff, Garrett, & John Von Neumann. (1936). The Logic of Quantum Mechanics. *Annals of Mathematics*, 37(4), 823–843. http://dx.doi.org/10.2307/1968621.

Bochvar, D. A. (1984). On the Consistency of a Three-Valued Logical Calculus. *Topoi*, 3(1), 3–12. http://dx.doi.org/10.1007/BF00136115.

Bolc, Leonard, & Piotr Borowik. (1992). *Many-Valued Logics 1: Theoretical Foundations*. Many-Valued Logics. Berlin, Heidelberg: Springer-Verlag. https://www.springer.com/us/book/9783540559269.

Boole, George. (2016). *The Mathematical Analysis of Logic: Being an Essay Towards a Calculus of Deductive Reasoning*. S.l. London: Forgotten Books.

Byrne, Peter. (2012). *The Many Worlds of Hugh Everett III: Multiple Universes, Mutual Assured Destruction, and the Meltdown of a Nuclear Family*. United Kingdom: OUP Oxford.

Church, Alonzo. (1939). Review: D. A. Bocvar, On a Three-Valued Logical Calculus and Its Application to the Analysis of Contradictions. *Journal of Symbolic Logic*, 4(2), 98–99.

Costa, Newton C. A. da. (1997). *Logiques classiques et non classiques: essai sur les fondements de la logique*. Paris: Masson.

Crease, Robert P. (2002). The Most Beautiful Experiment. *Physics World*, 15(9), 19. http://dx.doi.org/10.1088/2058-7058/15/9/22.

Destouches-Février, P. (1948). Logique et Théories Physiques. *Synthese*, 7(6-A), 400–410.

Dubois, François, & Zeno Toffano. (2017). Eigenlogic: A Quantum View for Multiple-Valued and Fuzzy Systems. In Jose Acacio de Barros, Bob Coecke, & Emmanuel Pothos edited by, *Quantum Interaction* (pp. 239–251). Lecture Notes in Computer Science. Switzerland: Springer International Publishing.

Einstein, A. (1905). Über Einen Die Erzeugung Und Verwandlung Des Lichtes Betreffenden Heuristischen Gesichtspunkt. *Annalen der Physik*, 322(6), 132–148. http://dx.doi.org/10.1002/andp.19053220607.

Epstein, G. (1993). *Multiple-Valued Logic Design: An Introduction*. United Kingdom: CRC Press.

Epstein, G., G. Frieder, & D. C. Rine. (1974). The Development of Multiple-Valued Logic as Related to Computer Science. *Computer*, 7(9), 20–32. http://dx.doi.org/10.1109/MC.1974.6323304.

Evans, James, & Alan S. Thorndike. (Eds.), (2007). *Quantum Mechanics at the Crossroads: New Perspectives from History, Philosophy and Physics*. The Frontiers Collection. Berlin, Heidelberg: Springer-Verlag. //www.springer.com/us/book/9783540326632.

Everett, Hugh. (2012). *The Everett Interpretation of Quantum Mechanics: Collected Works 1955-1980 with Commentary*. USA: Princeton University Press.

Fitting, Melvin, & Ewa Orlowska. (Eds.), (2003). *Beyond Two: Theory and Applications of Multiple-Valued Logic*. Studies in Fuzziness and Soft Computing. Berlin: Physica-Verlag Heidelberg. https://www.springer.com/gp/book/9783790815412.

Foucault, Léon. (1853). *Sur les vitesses relatives de la lumière dans l'air et dans l'eau*. 1. Paris: Impr. Bachelier. http://jubilotheque.upmc.fr/ead.html?id=TH_000075_001.

Fresnel, Augustin. (1868). *Oeuvres completes d'Augustin Fresnel Augustin Fresnel*. National Central Library of Florence. Imprimerie Imperiale. http://archive.org/details/bub_gb_MZyb3-HTc08C.

Gabbay, Dov M., Akihiro Kanamori, & John Woods. (Eds.), (2012). *Handbook of the History of Logic | Sets and Extensions in the Twentieth Century*. 1 edition. Vol. 6. Amsterdam: North Holland.

Gabbay, Dov M., Francis Jeffry Pelletier, & John Woods. (Eds.), (2012). *Handbook of the History of Logic | Logic: A History of Its Central Concepts*. 1 edition. Vol. 11. Amsterdam; Boston: North Holland.

Gabbay, Dov M., Jörg H. Siekmann, & John Woods. (Eds.), (2014). *Handbook of the History of Logic | Computational Logic*. 1 edition. Vol. 9. Amsterdam: North Holland.

Gabbay, Dov M., & John Woods. (Eds.), (2004a). *Handbook of the History of Logic | Greek, Indian and Arabic Logic*. 1 edition. Vol. 1. Amsterdam: North Holland.

Gabbay, Dov M., & John Woods. (Eds.), (2004b). *Handbook of the History of Logic | The Rise of Modern Logic: From Leibniz to Frege*. 1 edition. Vol. 3. Amsterdam; Boston: North Holland.

Gabbay, Dov M., & John Woods. (Eds.), (2006). *Handbook of the History of Logic | Logic and the Modalities in the Twentieth Century*. 1 edition. Vol. 7. Amsterdam: North Holland.

Gabbay, Dov M., & John Woods. (Eds.), (2008a). *Handbook of the History of Logic | British Logic in the Nineteenth Century*. 1 edition. Vol. 4. Amsterdam: North Holland.

Gabbay, Dov M., & John Woods. (Eds.), (2008b). *Handbook of the History of Logic | Mediaeval and Renaissance Logic*. 1 edition. Vol. 2. Amsterdam; Boston: North Holland.

Gabbay, Dov M., & John Woods. (Eds.), (2009). *Handbook of the History of Logic | Logic from Russell to Church*. 1 edition. Vol. 5. Amsterdam: North Holland.

Gabbay, Dov M., & John Woods. (Eds.), (2014). *Handbook of the History of Logic | The Many Valued and Nonmonotonic Turn in Logic*. 1 edition. Vol. 8. Amsterdam: North Holland.

Gabbay, Dov M., John Woods, & Stephan Hartmann. (Eds.), (2011). *Handbook of the History of Logic | Inductive Logic*. 1 edition. Vol. 10. Amsterdam: North Holland.

Gerasimov, A. S. (2018). Infinite-Valued First-Order Łukasiewicz Logic: Hypersequent Calculi Without Structural Rules and Proof Search for Sentences in the Prenex Form. *Siberian Advances in Mathematics*, 28(2), 79–100. http://dx.doi.org/10.3103/S1055134418020013.

Goguen, J. A. (1969). The Logic of Inexact Concepts. *Synthese*, 19(3), 325–373. http://dx.doi.org/10.1007/BF00485654.

Gonzalez, Wilmer Pereira. (1992). *Une logique modale pour le raisonnement dans l'incertain*. doctoral thesis, France: Rennes University.

Gourinat, Jean-Baptiste. (2011). Aristote et la « logique formelle moderne »: sur quelques paradoxes de l'interprétation de Łukasiewicz. *Philosophia Scientiæ. Travaux d'histoire et de philosophie des sciences*, 15–2(September), 69–101. http://dx.doi.org/10.4000/philosophiascientiae.654.

Gribbin, John. (2008). *Le chat de Schrödinger: physique quantique et réalité*. France: Alphée.

Hallden, S. (1949). *The Logic of Nonsense*. Sweden: Upsala Universitets Arsskrift.

Heelan, Patrick Aidan. (2015). *The Observable: Heisenberg's Philosophy of Quantum Mechanics*. New York: Peter Lang.

Heisenberg, Werner. (1962). *Physics & Philosophy: The Revolution in Modern Science*. New York: Harper & Row.

Hertz, H. (1887). Ueber Einen Einfluss Des Ultravioletten Lichtes Auf Die Electrische Entladung. *Annalen der Physik*, 267(8), 983–1000. http://dx.doi.org/10.1002/andp.18872670827.

Husson, Suzanne. (2009). *Interpréter le De interpretatione*. France: Vrin.

Huygens, Christian. (1678). *Traité de la lumière*. https://www.dunod.com/sciences-techniques/traite-lumiere-presente-par-michel-blay.

Jammer, Max. (1974). *The Philosophy of Quantum Mechanics: The Interpretations of Quantum Mechanics in Historical Perspective*. First Edition. New York: John Wiley & Sons Inc.

Jolivet, Jean. (1994). *Abélard, ou, La philosophie dans le langage*. Switzerland: Saint-Paul.
Kleene, S. C. (1938). On Notation for Ordinal Numbers. *The Journal of Symbolic Logic*, 3(4), 150–155. http://dx.doi.org/10.2307/2267778.
Kleene, Stephen Cole. (1952). *Introduction to Metamathematics*. New Jersey: Wolters-Noordhoff.
Lukasiewicz, J., L. Borkowski, & O. Wojtasiewicz. (1970). *Selected Works*. Amsterdam: North-Holland.
Łukasiewicz, Jan. (1920). O Logice Trójwartociowej. *Studia Filozoficzne*, 270(5), 170–171.
Łukasiewicz, Jan. (1930). Untersuchungen Über Den Aussagenkalkül. *Comptes Rendus de Séances de La Société Des Sciences et Des Lettres de Varsovie*, 3, 1–21.
Łukasiewicz, Jan. (2000). *Du principe de contradiction chez Aristote*. éditions de l'éclat.
Mamache, Fatiha. (2012). *Logique Temporelle*. France: PAF.
Matsumakia, Marc-Polycarpe Mutombo. (2010). *Précis de logique non classique*. France: Editions Publibook.
McNaughton, Robert. (1951). A Theorem About Infinite-Valued Sentential Logic. *The Journal of Symbolic Logic*, 16(1), 1–13. http://dx.doi.org/10.2307/2268660.
Miller, D. Michael, & Mitchell Aaron Thornton. (2008). *Multiple Valued Logic: Concepts and Representations*. USA: Morgan & Claypool Publishers.
Moussa, Soumaya, & Saoussen Bel Hadj Kacem. (2017). Symbolic Approximate Reasoning with Fuzzy and Multi-Valued Knowledge. *Procedia Computer Science*, Knowledge-Based and Intelligent Information & Engineering Systems: Proceedings of the 21st International Conference, KES-20176-8 September 2017, Marseille, France, 112 (January), 800–810. http://dx.doi.org/10.1016/j.procs.2017.08.048.
Newton, Isaac. (1704). *Opticks*. http://data.bnf.fr/15506384/isaac_newton_opticks/.
Piróg-Rzepecka, Krystyna. (1977). *Systemy nonsense-logics*. Poland: Państwowe Wydawn Naukowe.
Post, Emil L. (1921). Introduction to a General Theory of Elementary Propositions. *American Journal of Mathematics*, 43(3), 163–185. http://dx.doi.org/10.2307/2370324.
Putnam, Hilary. (1957). Three-Valued Logic. *Philosophical Studies: An International Journal for Philosophy in the Analytic Tradition*, 8(5), 73–80.
Pykacz, Jarosław. (2015). *Quantum Physics, Fuzzy Sets and Logic: Steps Towards a Many-Valued Interpretation of Quantum Mechanics*. Springer Briefs in Physics. Springer International Publishing. https://www.springer.com/la/book/9783319193830.
Raimond, Jean-Michel, & Michel Brune. (2013). Serge Haroche, prix Nobel de physique 2012. *La lettre du Collège de France*, 35 (December), 16–17. http://dx.doi.org/10.4000/lettre-cdf.2382.
Rasiowa, Helena. (1974). *An Algebraic Approach to Non-Classical Logics*. Amsterdam: North-Holland Publishing Company.
Reichenbach, Hans. (1946). Reply to Ernest Nagel's Criticism of My Views on Quantum Mechanics. *Journal of Philosophy*, 43(9), 239–247.
Rescher, Nicholas. (1968). Many-Valued Logic. In *Topics in Philosophical Logic* (pp. 54–125). Dordrecht: Synthese Library. Springer. http://dx.doi.org/10.1007/978-94-017-3546-9_6.
Rine, David C. (2014). *Computer Science and Multiple-Valued Logic: Theory and Applications*. Amsterdam: Elsevier Science.
Segerberg, Krister. (1965). A Contribution to Nonsense-Logics. *Theoria*, 31(3), 199–217.
Sokolov, Artem, & Oleg Zhdanov. (2019). Prospects for the Application of Many-Valued Logic Functions in Cryptography. In Zhengbing Hu, Sergey Petoukhov, Ivan Dychka, & Matthew He edited by, *Advances in Computer Science for Engineering and Education* (pp. 331–339). Advances in Intelligent Systems and Computing. Switzerland: Springer International Publishing.

Stanković, Radomir S., Jaakko Astola, & Claudio Moraga. (2012). *Representation of Multiple-Valued Logic Functions*. USA: Morgan & Claypool Publishers.

Szmielew, Wanda. (1946). Review: D. A. Bochvar, To the Question of Paradoxes of the Mathematical Logic and Theory of Sets. *Journal of Symbolic Logic*, 11(4), 129–129.

Tarski, Alfred. (1937). Bolesław Sobociński. Aksjomotyzacja Pewnych Wielowartościowych Systemów Teorji Dedukcji (Axiomatization of Certain Many-Valued Systems of the Theory of Deduction). Roczniki Prac Naukowych Zrzeszenia Asystentów Uniwersytetu Józefa Piłsudskiego w Warszawie, Vol. 1, Wydzial Matematyczno-Przyrodniczy Nr. 1, Warsaw1936, Pp. 399–419. *The Journal of Symbolic Logic*, 2(2), 93–93. http://dx.doi.org/10.2307/2267391.

Trimmer, John D. (1980). The Present Situation in Quantum Mechanics: A Translation of Schrödinger's 'Cat Paradox' Paper. *Proceedings of the American Philosophical Society*, 124(5), 323–338.

Turquette, Atwell R. (1963). Independent Axioms for Infinite-Valued Logic. *The Journal of Symbolic Logic*, 28(3), 217–221. http://dx.doi.org/10.2307/2271067.

Vranesic, Z. G., & K. C. Smith. (1974). Engineering Aspects of Multi-Valued Logic Systems. *Computer*, 7(9), 34–41. http://dx.doi.org/10.1109/MC.1974.6323306.

Whitehead, Alfred North, & Bertrand Russell. (1910). *Principia Mathematica*. Vol. 1. United Kingdom: Cambridge University Press.

Young, Thomas. (1802). The Bakerian Lecture. On the Theory of Light and Colours. *Philosophical Transactions of the Royal Society of London*, 92(January), 12–48. http://dx.doi.org/10.1098/rstl.1802.0004.

Zawirski, Sigismond. (1932). Les Logiques Nouvelles et Le Champ de Leur Application. *Revue de Métaphysique et de Morale*, 39(4), 503–519.

3 Fuzzy Logic

INTRODUCTION

"There is nothing fuzzy about fuzzy logic."

(Zimmermann, 2011)

Lotfi Zadeh is fully credited with fuzzy logic (Zadeh, 1965). However, like with many scientific disciplines, some authors laid bricks, others removed the defective ones, and some added the cement to bind the whole. This last is what Zadeh did, by linking logic and uncertainty in the fuzzy logic that will be detailed in this chapter.

According to Haack (1979), Zadeh offers us not only a radically nonclassical logic but also a radically nonstandard concept of the nature of logic. It would hardly be an exaggeration to say that fuzzy logic lacks all the features that pioneers of modern logic wanted from a logic. It is not only a logic of vagueness; it is, from Frege's point of view, what would have been a contradiction in terms, an oxymoron—a vague logic.

DEFINITIONS

Without pretending to be exhaustive, some necessary concepts are presented here (Bernadette, 2003; Sakawa, 1993).

Let X be a classical set and A a classical subset. In Boolean logic, the subset A is described by elements x in X that belong to A, and can be described by a characteristic function $\chi_A: X \to \{0, 1\}$ (Figure 3.1):

$$\chi_A(x) = \begin{cases} 1 & if\ x \in A \\ 0 & else. \end{cases} \quad (3.1)$$

In fuzzy logic, belonging is a matter of degree. Thus a fuzzy subset $\tilde{F}$ of X is defined not by a characteristic function but by the membership degree of its elements $\mu_{\tilde{F}}: X \to [0, 1]$:

$$\tilde{F} = \{<x, \mu_{\tilde{F}}(x) > | x \in X\}. \quad (3.2)$$

According to the nature of X, one can distinguish the finite case (Equation 3.3) from the infinite one (Equation 3.4):

$$\tilde{F} = \sum_{i=1}^{n} \mu_{\tilde{F}}(x_i)/x_i \in X, \quad (3.3)$$

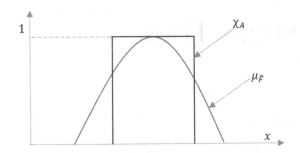

FIGURE 3.1 Membership Function and Characteristic Function.

$$\tilde{F} = \int_X \mu_{\tilde{F}}(x)/x \in X. \tag{3.4}$$

The fuzzy set support $Supp(\tilde{F})$ contains each element having nonnull membership:

$$Supp(\tilde{F}) = \{x \in X | \mu_{\tilde{F}}(x) > 0\}. \tag{3.6}$$

The height of a fuzzy set $h(\tilde{F})$ is the largest value of its membership function:

$$h(\tilde{F}) = \sup_{x \in X} \mu_{\tilde{F}}(x). \tag{3.7}$$

A fuzzy set is called normal if $h(\tilde{F}) = 1$.

A fuzzy set is convex if

$$\mu_{\tilde{F}}(\lambda x_1 + (1 - \lambda)x_2) \geq \min\left(\mu_{\tilde{F}}(x_1), \mu_{\tilde{F}}(x_2)\right), \tag{3.8}$$

$\forall\, x_1,\, x_2 \in X$ and $\lambda \in [0, 1]$.

A Fuzzy Number (FN) is a convex normalized fuzzy set whose membership function is piecewise continuous.

A Fuzzy Sub Number (FSN) is a FN for which the normality condition is not fulfilled (Javanmard & Nehi, 2018).

A FN $\tilde{N}$ is called positive if $\mu_{\tilde{N}}(x) = 0\ \forall\, x < 0$ verifying $x \in X$.

The most-used notation is based on Trapezoidal Fuzzy Numbers (TFNs), since they allow an acceptable compromise between representation simplicity and algorithmic complexity (Grzegorzewski & Mrówka, 2005, 2007). A TFN $\tilde{N}$ is described by four tuples $\tilde{N}\,(n^1, n^2, n^3, n^4)$ as follows (El Alaoui, 2020):

Fuzzy Logic

$$\mu_{\tilde{N}}(x) = \begin{cases} \frac{x-n^1}{n^2-n^1}, & n^1 \leq x \leq n^2 \\ 1, & n^2 \leq x \leq n^3 \\ \frac{x-n^4}{n^3-n^4}, & n^3 \leq x \leq n^4 \\ 0, & else \end{cases} \qquad (3.9)$$

If $n^2 = n^3$, the TFN is called a triangular fuzzy number. If $n^1 = n^2 = n^3 = n^4$, we return to classical crisp numbers.

Let $\tilde{A}$ and $\tilde{B}$ be two positive TFNs. Then fuzzy addition, subtraction, multiplication, multiplication by a scalar, and division are respectively defined by the following:

$$\tilde{A} \oplus \tilde{B} = (a^1 + b^1, a^2 + b^2, a^3 + b^3, a^4 + b^4) \qquad (3.10)$$

$$\tilde{A} \ominus \tilde{B} = (a^1 - b^4, a^2 - b^3, a^3 - b^2, a^4 - b^1) \qquad (3.11)$$

$$\tilde{A} \otimes \tilde{B} = (a^1 * b^1, a^2 * b^2, a^3 * b^3, a^4 * b^4) \qquad (3.12)$$

$$\beta \otimes \tilde{A} = (\beta * a^1, \beta * a^2, \beta * a^3, \beta * a^4) \qquad (3.13)$$

$$\tilde{A} \oslash \tilde{B} = (a^1/b^4, a^2/b^3, a^3/b^2, a^4/b^1). \qquad (3.14)$$

For arbitrary TFNs, not necessarily positive ones, these operations remain unchanged except for fuzzy multiplication (Equation 3.12), which becomes (Bansal, 2011)

$$\tilde{A} \otimes \tilde{B} = (e^1, e^2, e^3, e^4), \qquad (3.15)$$

verifying

$$e^1 = \min(a^1 * b^1, a^2 * b^2, a^3 * b^3, a^4 * b^4) \qquad (3.15.1)$$

$$e^2 = \min(a^2 * b^2, a^2 * b^3, a^3 * b^2, a^3 * b^3) \qquad (3.15.2)$$

$$e^3 = \max(a^2 * b^2, a^2 * b^3, a^3 * b^2, a^3 * b^3) \qquad (3.15.3)$$

$$e^4 = \max(a^1 * b^1, a^2 * b^2, a^3 * b^3, a^4 * b^4). \qquad (3.15.4)$$

A subnormal TFN (Figure 3.2) is called a generalized TFN $\tilde{N}(n^1, n^2, n^3, n^4; h^1, h^2)$:

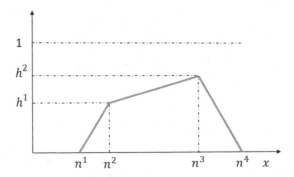

FIGURE 3.2 Generalized Trapezoidal Fuzzy Number.

$$\mu_{\tilde{N}}(x) = \begin{cases} h^1 \frac{x - n^1}{n^2 - n^1}, & n^1 \leq x \leq n^2 \\ (h^2 - h^1) \frac{x - n^2}{n^3 - n^2} + h^1, & n^2 \leq x \leq n^3 \\ h^2 \frac{x - n^4}{n^3 - n^4}, & n^3 \leq x \leq n^4 \\ 0, & otherwise. \end{cases} \quad (3.16)$$

Let $\tilde{A}(a^1, a^2, a^3, a^4, h^{1A}, h^{2A})$ and $\tilde{B}(b^1, b^2, b^3, b^4, h^{1B}, h^{2B})$ $\tilde{B}(b^1, b^2, b^3, b^4, h^{1B}, h^{2B})$ be two generalized TFNs. The height of the resulting generalized TFN $\tilde{C}(c^1, c^2, c^3, c^4, h^{1C}, h^{2C})$ using any of the fuzzy operations (Equations 3.10–3.15) is $h^{1C} = \min\{h^{1A}, h^{1B}\}$ and $h^{2C} = \min\{h^{2A}, h^{2B}\}$.

EXTENSIONS

How to estimate the membership function? Alternatively, how can we assign an exact membership function? While the main objective is to incorporate uncertainty, Zadeh (1971) admits that this is the weak point of fuzzy logic as proposed in 1965, which led him to introduce type 2 fuzzy sets (retroactively naming as type 1 fuzzy sets those he had already introduced; Zadeh, 1975). In type 2 fuzzy sets, the main idea is to take into account the uncertainty of the membership function—what can be described as a membership function of the membership function—giving rise to successive generalizations of fuzzy sets into type 3, type 4, and so on, through to type n, where for all $n \in IN^* \backslash \{1\}$, type n fuzzy sets are defined as fuzzy sets whose membership function takes values on the $n - 1$ fuzzy sets in the unit interval (Rickard, Aisbett, & Gibbon, 2009).

Several definitions have been proposed in the literature (Mendel & John, 2002; Mizumoto & Tanaka, 1976), with notations that are not always obvious to understand (Mendel, Rajati, & Sussner, 2016).

Type 2 fuzzy sets can be divided into two categories (Fazel Zarandi, Gamasaee, & Castillo, 2016):

- In generalized type 2 fuzzy sets, the second membership function can take all possible values.
- In Interval Type 2 Fuzzy Sets (IT2FSs), the second membership function is restricted to 0 and 1. We can see these also as type 2 fuzzy sets $\tilde{\tilde{F}}$ whose membership function is bounded by two type 1 fuzzy sets that constitute the upper bound $\tilde{F}^U$ and the lower bound $\tilde{F}^L$.

It is clear that generalized type 2 fuzzy sets are a generalization of IT2FSs. However, their handling requires considerable computational capacity (Liang & Mendel, 2000), and must be reserved for appropriate applications.

IT2FSs can be divided into three categories (El Alaoui, El Yassini, & Ben-azza, 2019; Javanmard & Nehi, 2018):

- An IT2FS $\tilde{\tilde{F}} = (\tilde{F}^L, \tilde{F}^U)$ is called a perfect Interval Type 2 FN (IT2FN) if $\tilde{F}^U$ and $\tilde{F}^L$ are FNs.
- An IT2FS $\tilde{\tilde{F}} = (\tilde{F}^L, \tilde{F}^U)$ is called an IT2FN if $\tilde{F}^U$ is a FN and $\tilde{F}^L$ is a FSN.
- An IT2FS $\tilde{\tilde{F}} = (\tilde{F}^L, \tilde{F}^U)$ is called a sub-IT2FN if $\tilde{F}^U$ and $\tilde{F}^L$ are FSNs.

According to Mendel (2017), only perfect IT2FNs deserve to be called IT2FNs.

Let $\tilde{\tilde{A}} = (\tilde{A}^L, \tilde{A}^U)$ and $\tilde{\tilde{B}} = (\tilde{B}^L, \tilde{B}^U)$ be two IT2FNs. The fuzzy arithmetic operations (Equations 3.10–3.14) are extended as follows:

$$\tilde{\tilde{A}} \oplus \tilde{\tilde{B}} = [\tilde{A}^L \oplus \tilde{B}^L, \tilde{A}^U \oplus \tilde{B}^U] \qquad (3.17)$$

$$\tilde{\tilde{A}} \ominus \tilde{\tilde{B}} = [\tilde{A}^L \ominus \tilde{B}^L, \tilde{A}^U \ominus \tilde{B}^U] \qquad (3.18)$$

$$\tilde{\tilde{A}} \otimes \tilde{\tilde{B}} = [\tilde{A}^L \otimes \tilde{B}^L, \tilde{A}^U \otimes \tilde{B}^U] \qquad (3.19)$$

$$\beta \otimes \tilde{\tilde{A}} = [\beta \otimes \tilde{A}^L, \beta \otimes \tilde{A}^U] \qquad (3.20)$$

$$\tilde{\tilde{A}} \oslash \tilde{\tilde{B}} = [\tilde{A}^L \oslash \tilde{B}^L, \tilde{A}^U \oslash \tilde{B}^U]. \qquad (3.21)$$

It should be noted that some authors consider IT2FSs identical to interval-valued fuzzy sets (Javanmard & Nehi, 2018), while others believe that IT2FSs are a generalization and that defined operations and methods are valid only when they coincide (Mendel et al., 2016). It should also be noted that the membership function is seen as being either a value belonging to the interval that cannot be determined accurately or the entire interval (Montero, Gómez, & Bustince, 2007).

Atanassov stipulates that in addition to the membership function, the non-membership function must be taken into consideration, which he illustrates by an electoral vote. Assume a population of 10 people voting for a project: five for, three against, and two abstaining. In this case, the membership function is $\mu_{\tilde{F}}(x) = 0.5$, the nonmembership function is $\vartheta_{\tilde{F}}(x) = 0.3$, and the hesitancy function—neither for nor against—is $\pi_{\tilde{F}}(x) = 0.2$. Hence the definition of intuitionistic fuzzy sets (Atanassov, 1986):

$$\tilde{F} = \{<x, \mu_{\tilde{F}}(x), \vartheta_{\tilde{F}}(x) > | x \in X\}. \qquad (3.22)$$

In the first definition, from 1986, membership, nonmembership, and hesitancy were singletons in the unit interval verifying

$$0 \leq \mu_{\tilde{F}}(x) \leq 1, \ 0 \leq \vartheta_{\tilde{F}}(x) \leq 1, \ 0 \leq \mu_{\tilde{F}}(x) + \vartheta_{\tilde{F}}(x) \leq 1 \text{ and}$$
$$\pi_{\tilde{F}}(x) = 1 - \left(\mu_{\tilde{F}}(x) + \vartheta_{\tilde{F}}(x)\right).$$

Similar to the development of type 1 fuzzy sets, and given the difficulty of assigning an exact value to the membership function, intuitionistic fuzzy sets were extended into interval-valued intuitionistic fuzzy sets (Atanassov & Gargov, 1989), in which the membership and nonmembership functions become no longer singletons but intervals:

$$\tilde{F} = \{<x, M_{\tilde{F}}(x), N_{\tilde{F}}(x) > | x \in X\} \qquad (3.23)$$

verifying $M_{\tilde{F}}(x) \subseteq [0, 1]$, $N_{\tilde{F}}(x) \subseteq [0, 1]$, and sup $M_{\tilde{F}}(x)$ + sup $N_{\tilde{F}}(x) \leq 1$.

An intuitionistic fuzzy set is convex if for all $x_1, x_2 \in X$ and $\lambda \in [0, 1]$,

$$\begin{aligned}\mu_{\tilde{F}}(\lambda x_1 + (1-\lambda)x_2) &\geq \min\left(\mu_{\tilde{F}}(x_1), \mu_{\tilde{F}}(x_2)\right) \text{ and} \\ \vartheta_{\tilde{F}}(\lambda x_1 + (1-\lambda)x_2) &\leq \min(\vartheta_{\tilde{F}}(x_1), \vartheta_{\tilde{F}}(x_2))\end{aligned} \qquad (3.24)$$

An intuitionistic fuzzy set is normalized if there is an x_0 verifying $\mu_{\tilde{F}}(x_0) = 1$ and $\vartheta_{\tilde{F}}(x_0) = 0$.

An intuitionistic fuzzy number is a convex and normalized intuitionistic fuzzy set.

A trapezoidal intuitionistic fuzzy number is described by eight tuples $(a^1, a^2, a^3, a^4; a^{1'}, a^{2'}, a^{3'}, a^{4'})$ as follows:

$$\mu_{\tilde{A}}(x) = \begin{cases} \frac{x-a^1}{a^2-a^1}, & a^1 \leq x \leq a^2 \\ 1, & a^2 \leq x \leq a^3 \\ \frac{x-a^4}{a^3-a^4}, & a^3 \leq x \leq a^4 \\ 0, & \text{else} \end{cases} \quad \vartheta_{\tilde{A}}(x) = \begin{cases} \frac{x-a^{1'}}{a^{2'}-a^{1'}}, & a^{1'} \leq x \leq a^{2'} \\ 0, & a^{2'} \leq x \leq a^{3'} \\ \frac{x-a^{3'}}{a^{3'}-a^{4'}}, & a^{3'} \leq x \leq a^{4'} \\ 0, & \text{else} \end{cases}, \qquad (3.25)$$

where $a^{1'} \leq a^1 \leq a^{2'} \leq a^2 \leq a^3 \leq a^{3'} \leq a^4 \leq a^{4'}$.

If $a^2 = a^3 = a^{2'} = a^{3'}$, the trapezoidal intuitionistic fuzzy number is called a triangular intuitionistic fuzzy number.

If $a^{1'} = a^{2'} = a^{3'} = a^{4'} = a^1 = a^2 = a^3 = a^4$, the intuitionistic fuzzy number is reduced to a real crisp number.

A review of the different types of fuzzy sets is presented by Bustince et al. (2016). The evolution of intuitionistic fuzzy sets is covered by Atanassov and Vassilev (2018).

RANKING

While it seems intuitive to say that the TFN (4, 5, 6, 7) is bigger than (1, 2, 2, 3), comparing (4, 5, 6, 7) and (4, 5.5, 5.5, 7) is trickier, since there is no complete order in fuzzy sets. Wang and Kerre (2001) advocate for seven reasonable properties for ranking FNs. Thus, given three FNs $\tilde{A}$, $\tilde{B}$, and $\tilde{C}$ (El Alaoui, Ben-azza, & Zahi, 2016) the following properties apply:

Property 1: $\tilde{A} \succcurlyeq \tilde{A}$, where $\succcurlyeq$ denotes the fuzzy version of $\geq$.
Property 2: If $\tilde{A} \succcurlyeq \tilde{B}$ and $\tilde{B} \succcurlyeq \tilde{C}$, then $\tilde{A} \succcurlyeq \tilde{C}$.
Property 3: If $\tilde{A} \succcurlyeq \tilde{B}$ and $\tilde{B} \succcurlyeq \tilde{A}$, then $\tilde{A} \cong \tilde{B}$, where $\cong$ denotes fuzzy equality.
Property 4: If $\tilde{A} \cap \tilde{B} = \emptyset$ and $\tilde{A}$ is to the right of $\tilde{B}$, then $\tilde{A} \succcurlyeq \tilde{B}$.
Property 5: The order of $\tilde{A}$ and $\tilde{B}$ is not affected by the other fuzzy numbers under comparison.
Property 6: If $\tilde{A} \succcurlyeq \tilde{B}$, then $\tilde{A} \oplus \tilde{C} \succcurlyeq \tilde{B} \oplus \tilde{C}$.
Property 7: If $\tilde{A} \succcurlyeq \tilde{B}$, then $\tilde{A} \otimes \tilde{C} \succcurlyeq \tilde{B} \otimes \tilde{C}$.

The different ranking methods are divided into three classes (Wang & Kerre, 2001). In the first, each fuzzy quantity being compared is linked to a real number through a ranking function $\mathcal{R}: \tilde{A} \to \mathcal{R}(\tilde{A})$. Thus, the comparison occurs between the associated crisp numbers. In the second, which is the closer to the spirit of TOPSIS, each fuzzy quantity is compared to a reference point. In the third, pairwise comparisons occur between each pair of fuzzy quantities (El Alaoui & El Yassini, 2020).

CONCLUSION

This chapter provided a brief overview on the development of fuzzy logic, beginning by differentiating fuzzy sets from classical sets and then introducing fuzzy numbers and necessary operations by presenting the most-used forms and appropriate mathematical operators. Then starting from the difficulty of estimating the value of the membership function, generalizations into type 2 fuzzy sets and intuitionistic fuzzy logic were presented, specifying the differences among the extensions and their associated mathematical background. Some differences in approach between classical logic and fuzzy logic were highlighted in the chapter. The chapter also pointed out some ranking approaches for fuzzy quantities.

REFERENCES

Atanassov, Krassimir T. (1986). "Intuitionistic Fuzzy Sets". *Fuzzy Sets and Systems*, 20(1), 87–96. https://doi.org/10.1016/S0165-0114(86)80034-3.

Atanassov, Krassimir T., and G. Gargov (1989). "Interval Valued Intuitionistic Fuzzy Sets". *Fuzzy Sets and Systems*, 31(3), 343–349. https://doi.org/10.1016/0165-0114(89)90205-4.

Atanassov, Krassimir T., and Peter Vassilev (2018). "On the Intuitionistic Fuzzy Sets of N-Th Type". In *Advances in Data Analysis with Computational Intelligence Methods: Dedicated to Professor Jacek Żurada*, edited by Adam E. Gawęda, Janusz Kacprzyk, Leszek Rutkowski, and Gary G. Yen, (pp. 265–274). Studies in Computational Intelligence. Cham: Springer International Publishing. https://doi.org/10.1007/978-3-319-67946-4_10.

Bansal A. (2011). "Trapezoidal Fuzzy Numbers (a, b, c, d): Arithmetic Behavior". *International Journal of Physical and Mathematical Sciences*, 2(1), 39–44.

Bernadette, Bouchon-Meunier (2003). *Logique floue,principes, aide à la décision*. Paris: Lavoisier.

Bustince, H., E. Barrenechea, M. Pagola, J. Fernandez, Z. Xu, B. Bedregal, J. Montero, H. Hagras, F. Herrera, and B. De Baets (2016). "A Historical Account of Types of Fuzzy Sets and Their Relationships". *IEEE Transactions on Fuzzy Systems*, 24(1), 179–194. https://doi.org/10.1109/TFUZZ.2015.2451692.

El Alaoui, Mohamed (2020). "Fuzzy Goal Programming for Biodiesel Production". *International Journal of Green Energy*, 17(10), 1–8. https://doi.org/10.1080/15435075.2020.1779075.

El Alaoui, Mohamed, Hussain Ben-azza, and Azeddine Zahi (2016). "New Multi-Criteria Decision-Making Based on Fuzzy Similarity, Distance and Ranking". In *Proceedings of the Third International Afro-European Conference for Industrial Advancement—AECIA 2016*, (pp. 138–148). Advances in Intelligent Systems and Computing. Cham: Springer. https://doi.org/10.1007/978-3-319-60834-1_15.

El Alaoui, Mohamed, and Khalid El Yassini (2020). "Fuzzy Similarity Relations in Decision Making". *Handbook of Research on Emerging Applications of Fuzzy Algebraic Structures*, 369–385. https://doi.org/10.4018/978-1-7998-0190-0.ch020.

El Alaoui, Mohamed, Khalid El Yassini, and Hussain Ben-azza (2019). "Type 2 Fuzzy TOPSIS for Agriculture MCDM Problems". *International Journal of Sustainable Agricultural Management and Informatics*, 5(2/3), 112–130. https://doi.org/10.1504/IJSAMI.2019.101672.

Fazel Zarandi, M. H., R. Gamasaee, and O. Castillo (2016). "Type-1 to Type-n Fuzzy Logic and Systems". In *Fuzzy Logic in Its 50th Year: New Developments, Directions and Challenges*, edited by Cengiz Kahraman, Uzay Kaymak, and Adnan Yazici, (pp. 129–157). Studies in Fuzziness and Soft Computing. Cham: Springer International Publishing. https://doi.org/10.1007/978-3-319-31093-0_6.

Grzegorzewski, Przemysław, and Edyta Mrówka (2005). "Trapezoidal Approximations of Fuzzy Numbers". *Fuzzy Sets and Systems*, 153(1), 115–135. https://doi.org/10.1016/j.fss.2004.02.015.

Grzegorzewski, Przemysław, and Edyta Mrówka (2007). "Trapezoidal Approximations of Fuzzy Numbers—Revisited". *Fuzzy Sets and Systems*, 158(7), 757–768. https://doi.org/10.1016/j.fss.2006.11.015.

Haack, Susan (1979). "Do We Need 'Fuzzy Logic'?" *International Journal of Man-Machine Studies*, 11(4), 437–445. https://doi.org/10.1016/S0020-7373(79)80036-X.

Javanmard, Moslem, and Hassan Mishmast Nehi (2018). "Rankings and Operations for Interval Type-2 Fuzzy Numbers: A Review and Some New Methods". *Journal of Applied Mathematics and Computing*, May. https://doi.org/10.1007/s12190-018-1193-9.

Liang, Qilian, and J. M. Mendel (2000). "Interval Type-2 Fuzzy Logic Systems: Theory and Design". *IEEE Transactions on Fuzzy Systems*, *8*(5), 535–550. https://doi.org/10.1109/91.873577.

Mendel, J. M., H. Hagras, H. Bustince, and F. Herrera (2016). "Comments on 'Interval Type-2 Fuzzy Sets Are Generalization of Interval-Valued Fuzzy Sets: Towards a Wide View on Their Relationship'". *IEEE Transactions on Fuzzy Systems*, *24*(1), 249–250. https://doi.org/10.1109/TFUZZ.2015.2446508.

Mendel, J. M., and R. I. B. John (2002). "Type-2 Fuzzy Sets Made Simple". *IEEE Transactions on Fuzzy Systems*, *10*(2), 117–127. https://doi.org/10.1109/91.995115.

Mendel, Jerry M. (2017). *Uncertain Rule-Based Fuzzy Systems: Introduction and New Directions*. 2nd edition. Switzerland: Springer International Publishing. www.springer.com/us/book/9783319513690.

Mendel, Jerry M., Mohammad R. Rajati, and Peter Sussner (2016). "On Clarifying Some Definitions and Notations Used for Type-2 Fuzzy Sets as Well as Some Recommended Changes". *Information Sciences* C (340–341), 337–345. https://doi.org/10.1016/j.ins.2016.01.015.

Mizumoto, Masaharu, and Kokichi Tanaka (1976). "Some Properties of Fuzzy Sets of Type 2". *Information and Control*, *31*(4), 312–340. https://doi.org/10.1016/S0019-9958(76)80011-3.

Montero, J., D. Gómez, and H. Bustince (2007). "On the Relevance of Some Families of Fuzzy Sets". *Fuzzy Sets and Systems*, Theme: Basic Notions, *158*(22), 2429–2442. https://doi.org/10.1016/j.fss.2007.04.021.

Rickard, J. T., J. Aisbett, and G. Gibbon (2009). "Fuzzy Subsethood for Fuzzy Sets of Type-2 and Generalized Type-n". *IEEE Transactions on Fuzzy Systems*, *17*(1), 50–60. https://doi.org/10.1109/TFUZZ.2008.2006369.

Sakawa, Masatoshi (1993). *Fuzzy Sets and Interactive Multiobjective Optimization*. New York: Springer Science & Business Media.

Wang, Xuzhu, and Etienne E. Kerre (2001). "Reasonable Properties for the Ordering of Fuzzy Quantities (I)". *Fuzzy Sets and Systems*, *118*(3), 375–385. https://doi.org/10.1016/S0165-0114(99)00062-7.

Zadeh, L. A. (1965). "Fuzzy Sets". *Information and Control*, *8*(3), 338–353. https://doi.org/10.1016/S0019-9958(65)90241-X.

Zadeh, L. A. (1971). "Quantitative Fuzzy Semantics." *Information Sciences*, *3*(2), 159–176. https://doi.org/10.1016/S0020-0255(71)80004-X.

Zadeh, L. A. (1975). "The Concept of a Linguistic Variable and Its Application to Approximate Reasoning—I". *Information Sciences*, *8*(3), 199–249. https://doi.org/10.1016/0020-0255(75)90036-5.

Zimmermann, Hans-Jürgen (2011). *Fuzzy Set Theory—and Its Applications*. New York: Springer Science & Business Media.

4 Frequently Used Multicriteria Decision-Making Methods

INTRODUCTION

Multicriteria decision making—also referred to as multicriteria decision analysis—as a part of operations research, consists of evaluating and ranking solutions in order to find the best one. The process takes into consideration conflicting criteria and decision-maker preferences (El Alaoui, Ben-azza, & El Yassini, 2019). It can be divided into two main categories: multiobjective decision making and multiattribute decision making. In the former, a certain objective function (or objective functions) is (or are) to be optimized, dealing with an infinite number of possibilities on a continuous decision space. In the latter, the choice is between a restricted set of alternatives, and hence the notion of the best one is raised (El Alaoui, El Yassini, & Ben-azza, 2019).

It must be noted that a single-criterion problem is only an oversimplification of reality. Reducing everything to the economic aspect is just a grossly biased representation. Farmers, for example, are interested not only in maximizing their profit but also in minimizing environmental impacts, among other objectives (Romero & Rehman, 2003).

As soon as we recognize the multiplicity of criteria, we recognize also their conflicting nature. There might not be an optimal solution satisfying all criteria simultaneously, which implies a compromise solution (Zelany, 1974). A compromise solution, also called a feasible or nondominated solution, means that there is no other solution that outperforms it in accordance with all criteria (Clímaco, Mello, & Meza, 2008). Further discussions can be found with regard to the fuzzy environment (Li & Lai, 2000; Opricovic, 2007; Vahdani et al., 2013; Vinodh, Sarangan, & Chandra Vinoth, 2014) and the fully fuzzy framework (Hamadameen & Hassan, 2018), and linked to TOPSIS (Baky & Abo-Sinna, 2013; Dey, Pramanik, & Giri, 2014; Lai, Liu, & Hwang, 1994; Opricovic & Tzeng, 2004; Tzeng & Huang, 2016; Yoon, 1987).

According to Mukherjee (2017), the decision-making process contains, in most cases, eight steps:

- Step 1. Define the problem. A decision-making problem starts with defining the problem to be dealt with, its root causes, and its constraints or limitations.
- Step 2. Identify needs. Explain the need in detail to identify the feasible solution space.

- Step 3. Define goals. Specify the objectives to be solved in order to resolve the problem.
- Step 4. Select alternatives. Characterize all possible alternatives.
- Step 5. Identify suitable criteria.
- Step 6. Select an appropriate decision-making tool. The best-suited method is the least cumbersome and most error-free.
- Step 7. Evaluate all alternatives against the criteria.
- Step 8. Validate the result. In order to avoid any disagreement, the feasibility and acceptability of the result obtained must be checked prior to using it.

In a more concise vision, the approach proposed by Opricovic and Tzeng (2004) assumes that the desired goals have already been defined. There are thus six steps:

- Step 1. Define adequate criteria.
- Step 2. Identify suitable alternatives.
- Step 3. Evaluate each alternative in accordance to each criterion.
- Step 4. Use an adequate multicriteria analysis tool or techniques.
- Step 5. Accept the chosen alternative (preferable).
- Step 6. If the chosen solution is not feasible or acceptable, then collect more information and opt for next iteration.

Focusing on multiattribute decision-making methods—especially the analytic hierarchy process—Kubler et al. (2016) advocate for the following six steps:

- Step 1. Identify the aim of the process.
- Step 2. Identify alternatives.
- Step 3. Identify criteria.
- Step 4. Analyze alternatives.
- Step 5. Make choices.
- Step 6. Analyze feedback.

A similar approach excluding the last step is given by Rogers, Bruen, and Maystre (2000).

It is true that there are still disagreements over the steps to be taken. However, a general pattern seems to have been drawn up. Taking into consideration the complexity of the issues dealt with, spanning several strands and a multiplicity of criteria that no decision maker can deal with alone, the process involves several decision makers. Thus, a multiattribute decision-making problem can be formulated as follows:

A set of m alternatives A_i: $(1 \leq i \leq m)$ are to be evaluated by K decision makers D_k: $(1 \leq k \leq K)$ in accordance with n criteria C_j: $(1 \leq j \leq n)$.

Even if all the steps are important, some are more crucial than others. The choice of criteria, however important, is not a guarantee of the outcome of the process. Moreover, in comparing two alternatives according to a given criterion, the result can be one of the following (Teghem, Delhaye, & Kunsch, 1989):

- Indifferent, where neither alternative can be preferred over the other in accordance with the criterion
- Strictly preferred
- Weakly preferred
- Incomparable with regard to the criterion, due to uncertainty (Mukherjee, 2017)

A similar problem to weak preference is mentioned by Poincaré (2017): having a weight of 10 grams in your hand feels much like lifting 11 grams. A comparable similarity occurs with 11 and 12 grams. However, 10 and 12 grams seem to be more discernable from each other. This can be notated as 10 grams ~11 grams and 11 grams ~12 grams but 10 grams < 12 grams (Schärlig, 1985). This confirms the vagueness discussed in Chapter 1.

Is there an absolute best method? Or put another way, which is the best method for a specific problem?

Despite the abundance of methods to address the issue (Guitouni & Martel, 1998; Roy & Słowiński, 2013), the search for an absolute best method seems meaningless (Dymova, Sevastjanov, & Tikhonenko, 2016), since no single method can outshine all others in all aspects of all problems (Triantaphyllou, 2000). Thus, the problem becomes one of choosing the most suitable method for a certain problem. Some research trials to find the most appropriate method have been proposed in natural-resource management (Mendoza & Martins, 2006), renewable energy development (Kurka & Blackwood, 2013), and warehouse selection (Özcan, Çelebi, & Esnaf, 2011).

After tuning down our expectations from multiattribute decision-making methods, the remainder of this chapter presents the most used ones.

WEIGHTED SUM AND WEIGHTED PRODUCT

The most intuitive way to aggregate data is based on the weighted sum

$$WS_i = \sum_j w_j * y_{ij},$$

where WS_i is the weighted sum of the ith alternative, y_{ij} are the aggregated quantities representing the performance of the ith alternative in accordance with the jth criterion, and w_j are their weights. Generally, $\sum_j w_j = 1$.

Under the assumption of *homo economicus*, meaning a rational decision maker that desires to maximize its satisfaction (Guitouni & Martel, 1998), the alternative chosen is the one that optimizes (minimizes or maximizes) the weighted sum.

Despite its computational ease, the weighted sum suffers from many limitations. First, it requires that the entry elements be both numerical and comparable. Second, its compensatory nature may not be desirable; a good performance in accordance with a certain criterion with a relatively high weight can mask other underperformances while still producing a satisfactory result (El Alaoui & Ben-azza, 2017a).

Multiplicative aggregation overcomes some of the disadvantages of the additive approach, even if it is also compensatory. It permits the elimination of alternatives with poor attributes by forcing a simultaneous consideration of all criteria:

$$WP_i = \prod_j y_{ij}^{w_j}.$$

Limits and extensions of multiplicative aggregation in the fuzzy context have been proposed by El Alaoui and Ben-azza (2017b).

ANALYTIC HIERARCHY PROCESS

The Analytic Hierarchy Process (AHP), introduced by Saaty (1977, 1980), is concerned not with the performance of a specific alternative according to a given criterion but with the result of the comparison between that alternative and the others. This method is particularly interesting when the decision maker is more concerned with ranking than with performance itself (Papathanasiou et al., 2016). Moreover, in an environment punctuated by change, comparison is easier than evaluation. Thus, the method rests on pairwise comparisons between criteria and between alternatives in accordance with selection criteria, which give rise to a pairwise comparison matrix as follows:

$$A_{m \times m} = (a_{lc})_{m \times m} = \begin{bmatrix} a_{11} & \cdots & a_{1m} \\ \vdots & \ddots & \vdots \\ a_{m1} & \cdots & a_{mm} \end{bmatrix}.$$

The matrix size is $m \times m$ when comparing alternatives in accordance with criteria, or when comparing criteria themselves. It is frequently based on the 1–9 scale in Table 4.1, proposed by Saaty. The matrix must be both of the following:

- Reciprocal: $a_{lc} = a_{cl}^{-1}$, which imposes that $a_{ll} = 1$.
- Transitive: if alternative A_1 is preferred to A_2, and A_2 is preferred to A_3, then A_1 must be preferred to A_3 satisfying $a_{lc} = a_{lo} * a_{oc}$. However, if A_1 is four times preferred to A_2 and A_2 is three times preferred to A_3, trying to satisfy the equation indicates that A_1 will be 12 times preferred to A_3, which does not exist in the comparative scale. However, we must distinguish the numbers from their interpretations and the reality they try to represent, especially when the numbers are the result of a calculation. If a personal computer PC1 performs four times faster than PC2, and PC2 preforms three times faster than PC3, we can deduce that PC1 performs 12 times faster than PC3. Yet if PC1 costs four dollars more than PC2 and PC2 costs three dollars more than PC3, the deduction is additive, not multiplicative (Saaty, 2016).

TABLE 4.1
AHP Scale (Kunasekaran & Krishnamoorthy, 2014)

Definition	Numerical Rating	Reciprocal
Equal preference	1	1
Weak preference	2	1/2
Moderate preference	3	1/3
Moderate to strong preference	4	1/4
Strong preference	5	1/5
Strong to very strong preference	6	1/6
Very strong preference	7	1/7
Very strong to extreme preference	8	1/8
Extreme preference	9	1/9

This requires a consistency verification that occurs through the following steps:

- Compute the eigenvalues of the matrix. The maximal one will be noted λ_{max}.
- Compute the consistency ratio

$$CR = \frac{\lambda_{max} - m}{(m-1)RI_m},$$

where m represents the number of elements compared and RI_m the random index deduced from Table 4.2.

If $CR \leq 0.1$, the matrix is considered consistent; otherwise it must be reviewed.

When the comparison matrix is consistent, compute attribute weights w_{ij} in accordance with criteria j:

$$w_{ij} = \frac{\sum_{l=1}^{m} x_{lc}}{\sum_{c=1}^{m} \sum_{l=1}^{m} x_{lc}}.$$

In the same way, exchanging n for m, compute the criteria weight w_j.

TABLE 4.2
Random Index (Mukherjee, 2014)

m	1, 2	3	4	5	6	7	8	9	10	11	12	13	14
RI_m	0	0.58	0.9	1.12	1.24	1.32	1.41	1.45	1.49	1.52	1.54	1.56	1.58

TABLE 4.3
Pairwise Evaluation Matrix for Criteria and Alternatives

Criteria	C_1	C_2	C_3	Alternatives	C_1			C_2			C_3		
					A_1	A_2	A_3	A_1	A_2	A_3	A_1	A_2	A_3
C_1	1	2	3	A_1	1	1	1	1	3	2	1	1/9	1/8
C_2	1/2	1	2	A_2	1	1	1	1/3	1	1/2	9	1	1
C_3	1/3	1/2	1	A_3	1	1	1	1/2	2	1	8	1	1

The alternative with the highest weighted sum will be ranked first:

$$w_i = \sum_{j=1}^{n} w_{ij} * w_j.$$

However, some have questioned this method's usefulness, since it can produce counterintuitive results. In the example treated by Asadabadi, Chang, and Saberi (2019), containing three criteria (C_1, C_2, and C_3) and three alternatives (A_1, A_2, and A_3), all alternatives are equally preferred in accordance with C_1. A_1 is slightly preferred in accordance with C_2, and is largely dominated by the other alternatives (A_2 and A_3) in accordance with C_3. Table 4.3 details the comparisons.

Even if C_2 is considered more important than C_3, A_1 is expected last, because A_2 and A_3 are so much more highly favored in accordance with C_3 than A_1 is in accordance with C_2. However, AHP ranks A_1 first, which appears counterintuitive. It should be noted that the consistency ratio is acceptable in all four matrices.

Deeper discussions have been published considering AHP's criteria (Russo & Camanho, 2015), subcriteria (Mu & Pereyra-Rojas, 2017), and evolution (Emrouznejad & Marra, 2017), in addition to consistency (Aguarón et al., 2019) and consensus measurement (Dong, Fan, & Yu, 2015). A recent review was provided by Kubler et al. (2016) and dedicated books have been published (Bhushan & Rai, 2004; Brunelli, 2015; Emrouznejad, Ho, & Ho, 2017; Golden, Wasil, & Harker, 1989; Kou et al., 2013; Mu & Pereyra-Rojas, 2018; Saaty & Vargas, 2006, 2012).

ELECTRE

Assuming that no alternative dominates the others—which justifies the very existence of multiattribute decision-making methods—the family of ELECTRE methods (ELimination Et Choix Traduisant la REalité [Elimination and Choice Translating Reality]; Greco, Ehrgott, & Figueira, 2016) tries to make the decision maker accept, at a certain risk, the choice of one alternative over another. Versions of ELECTRE differ according to the required information, or the degree of

Frequently Used MCDM Methods

complexity (Belton & Stewart, 2002). The initial version (Roy, 1968) was proposed in order to overcome the shortcomings of the weighted sum (Greco et al., 2016), based on concordance and discordance computation. Having an evaluation matrix for each alternative according to each criterion and criterion weight. The concordance between two alternatives is defined by

$$C(A_1, A_2) = \frac{\sum_{l \in Q(A_1, A_2)} w_j}{\sum_j^n w_j}, \quad (4.1)$$

where $Q(A_1, A_2)$ represents the set of criteria for which A_1 is preferred to A_2
Discordance is computed by

$$D(A_{i_1}, A_{i_2}) = \frac{\max_{l \in R(A_{i_1}, A_{i_2})} w_j(c_j(A_2) - c_j(A_1))}{\max_{1 \leq j \leq m} \max_{A_{i1}, A_{i2} \in A} w_j(c_j(A_{i1}) - c_j(A_{i2}))}, \quad (4.2)$$

where $R(A_1, A_2)$ is the set of criteria for which A_2 is strictly preferred to A_1 and A is the set of alternatives.

Hence A_1 is preferred to A_2 if the concordance is superior to a predetermined value $\hat{c}$, the concordance threshold, such that $C(A_1, A_2) \geq \hat{c}$; and the discordance is inferior to another predetermined value $\hat{d}$, the discordance threshold, such that $D(A_1, A_2) \leq \hat{d}$ (Rogers et al., 2000).

In Table 4.4, four alternatives (A_1, A_2, A_3, and A_4) are evaluated according to four criteria (C_1, C_2, C_3, and C_4). All criteria are to be maximized.

The concordance between A_1 and A_2 is the sum of criterion weights according to which A_1 is better than A_2 divided by the sum of the weights (Equation 4.1):

$$C(A_1, A_2) = \frac{0.35}{0.35 + 0.1 + 0.3 + 0.25} = 0.35.$$

TABLE 4.4
ELECTRE Example

Assessment		Criterion			
		C_1	C_2	C_3	C_4
Alternative	A_1	1	0	0	0.77
	A_2	0	0.33	0.142	1
	A_3	0.4	1	1	0
	A_4	0.6	0.66	0.429	0.55
Criterion weight		0.35	0.1	0.3	0.25

TABLE 4.5
Concordance between Alternatives

Concordance			Alternative			
			A_1	A_2	A_3	A_4
Alternative		A_1	—	0.35	0.6	0.6
		A_2	0.65	—	0.25	0.25
		A_3	0.4	0.75	—	0.4
		A_4	0.4	0.75	0.6	—

Table 4.5 details the concordance measures between alternatives.

The discordance is the maximal discordance between the alternative being considered in accordance with criteria divided by the maximal discordance between each pair of alternatives; in this case it equals 1.

Thus the discordance between A_1 and A_2—is (Equation 4.2)

$$D(A_1, A_2) = \frac{\max((0.33 - 0), (0.142 - 0), (1 - 0.77))}{1} = 0.33.$$

Table 4.6 details the discordances between each pair of alternatives.

Fixing $\hat{c} = 0.4$, the pairs of alternatives satisfying $C(A_{i1}, A_{i2}) \geq 0.4$ are

(A_1, A_4), (A_1, A_3), (A_2, A_1), (A_3, A_1), (A_3, A_2), (A_3, A_4), (A_4, A_1), (A_4, A_2), (A_4, A_3).

Fixing $\hat{d} = 0.6$, the pairs of alternatives satisfying $C(A_{i1}, A_{i2}) \leq 0.6$ are

(A_1, A_2), (A_2, A_4), (A_3, A_4), (A_4, A_1), (A_4, A_2), (A_4, A_3).

TABLE 4.6
Discordance between Alternatives

Discordance			Alternative			
			A_1	A_2	A_3	A_4
Alternative		A_1	—	0.33	1	0.66
		A_2	1	—	0.858	0.6
		A_3	0.77	1	—	0.55
		A_4	0.4	0.45	0.51	—

Thus the ranking preferences are $(A_3 \geq A_4)$, $(A_4 \geq A_1)$, $(A_4 \geq A_2)$, $(A_4 \geq A_3)$.

From this, we can deduce that the chosen alternatives are A_3 and A_4, with indifferent preference, and that A_1 and A_2 are outranked by A_4.

While ELECTRE I is limited to elaborating preferences between nondominated solutions, further extensions (Table 4.7) are applicable for sorting, selection, and classification (Rogers et al., 2000). In that sense, ELECTRE II encompasses a complete ranking of alternatives from best to worst (Roy & Bertier, 1973). Recognizing the inherent uncertainty of the problems dealt with and the limits of probabilities, fuzzy logic has been incorporated into ELECTRE III (Roy, 1977), linked at the time to weak preferences.

ELECTRE IV (Roy & Hugonnard, 1982) may be considered the naive one; while it is pretty close to ELECTRE III, it supposes that all criteria have the same importance, which is more than debatable.

ELECTRE IS, like ELECTRE I, is a selection procedure that permits identification of incomparable alternatives. It may be considered an adaptation of ELECTRE I to fuzzy logic (Rogers et al., 2000). ELECTRE TRI (Roy & Bouyssou, 1993) assigns alternatives to predefined categories. ELECTRE successive extensions were presented in:

ELECTRE I: (Roy, 1968)
ELECTRE II: (Roy & Bertier, 1973)
ELECTRE III: (Roy, 1977)
ELECTRE IV: (Roy & Hugonnard, 1982)
ELECTRE IS: (Roy & Skalka, 1985)
ELECTRE TRI: (Roy & Bouyssou, 1993)

Is there a best ELECTRE version? Similar to finding the absolute best method, Rogers et al. (2000) insist that there is no best ELECTRE version, only versions better suited than others to some situations. Thus, the choice of the appropriate version must take into account the existence or absence of fuzziness and ultimately the purpose of the procedure.

Recent reviews of the ELECTRE methodology and its evolution are presented by Figueira et al. (2013), Figueira, Mousseau, and Roy (2016), and Govindan and Jepsen (2016), and related books have been published by Maystre, Pictet, and Simos (1994), Rogers et al. (2000), and Schärlig (1996).

TABLE 4.7
Choice of the Appropriate ELECTRE Version

Data	Aim		
	Selection	Allocation	Ranking
Crisp	I	—	II
Fuzzy	IS	TRI	III, IV

PROMETHEE

Preference Ranking Organization METHod for Enrichment of Evaluations (PROMETHEE; Mareschal, Brans, & Vincke, 1984), like ELECTRE, is an outranking method. It has also evolved into several versions, including the following (Chen, 2014)

- PROMETHEE I for partial ranking (Brans, 1982)
- PROMETHEE II for complete ranking (Brans, 1982)
- PROMETHEE III for ranking based on intervals (Mareschal et al., 1984)
- PROMETHEE IV for continuous cases (Mareschal et al., 1984)
- PROMETHEE GAIA for geometric analysis (Brans & Mareschal, 2001; Brans & Marshall, 1994; Mareschal & Brans, 1988)
- PROMETHEE V for net flows and integer linear programming (Brans & Mareschal, 1992)
- PROMETHEE VI for representation of the human brain (Brans & Mareschal, 1995)
- PROMETHEE GDSS for a group decision support system (Mareschal, Brans, & Macharis, 1998)
- PROMETHEE TRI for dealing with sorting problems (Figueira, De Smet, & Brans, 2005)
- PROMETHEE CLUSTER for nominal classification (Figueira et al., 2005)
- PROMETHEE GKS for robust ordinal regression (Kadziński, Greco, & Słowiński, 2012)

The method requirements are as follows (Cristobal, 2012):

- Evaluation of alternatives in accordance with each criterion $f_j(A_i)$, where i ($1 \le i \le m$) describes alternatives and j ($1 \le j \le n$) describes criteria, generating values
- Criterion weights w_j; in general, $\sum w_j = 1$

Criterion types specified as detailed in Table 4.8.

Since PROMETHEE is based on pairwise comparisons, it starts with computing the distances between each pair of alternatives in accordance with each criterion:

$$d_j(f_j(A_{i1}), f_j(A_{i2})) = f_j(A_{i1}) - f_j(A_{i2}). \tag{4.3}$$

To simplify, $d_j(f_j(A_{i1}), f_j(A_{i2}))$ will be denoted $d_j(A_{i1}, A_{i2})$.

Hence, based on the desired criterion type (Table 4.8), preference functions are computed by

$$P_j(A_{i1}, A_{i2}) = F_j(d_j(A_{i1}, A_{i2})) \tag{4.4}$$

TABLE 4.8
PROMETHEE Criterion Types

Type	Function	Parameters to Fix
Usual	$P_j(d_j) = \begin{cases} 0 & \text{if } d_j \leq 0 \\ 1 & \text{else} \end{cases}$	—
U-shape	$P_j(d_j) = \begin{cases} 0 & \text{if } d_j \leq q_j \\ 1 & \text{else} \end{cases}$	q_j
V-shape	$P_j(d_j) = \begin{cases} \frac{|d_j|}{p_j} & \text{if } d_j \leq p_j \\ 1 & \text{else} \end{cases}$	p_j
Level	$P_j(d_j) = \begin{cases} 0 & \text{if } |d_j| \leq q_j \\ 0.5 & \text{if } q_j < |d_j| \leq p_j \\ 1 & \text{else} \end{cases}$	p_j, q_j
Linear	$P_j(d_j) = \begin{cases} 0 & \text{if } |d_j| \leq q_j \\ \frac{|d_j| - q_j}{p_j - q_j} & \text{if } q_j < |d_j| \leq p_j \\ 1 & \text{else} \end{cases}$	p_j, q_j
Gaussian	$P_j(d_j) = 1 - e^{-\frac{d_j^2}{s_j^2}}$	s_j

satisfying $0 \leq P_j(A_{i1}, A_{i2}) \leq 1$.

For the cost criterion (minimization), the preference function is computed as

$$P_j(A_{i1}, A_{i2}) = F_j(-d_j(A_{i1}, A_{i2})). \tag{4.5}$$

Thus the preference degree of alternative A_{i1} relative to alternative A_{i2} is

$$\pi(A_{i1}, A_{i2}) = \sum_{j=1}^{n} P_j(A_{i1}, A_{i2}) * w_j. \tag{4.6}$$

Similarly the preference of A_{i2} over A_{i1} is

$$\pi(A_{i2}, A_{i1}) = \sum_{j=1}^{n} P_j(A_{i2}, A_{i1}) * w_j. \tag{4.7}$$

It should be noted that $0 \leq \pi(A_{i1}, A_{i2}) \leq 1$, $0 \leq \pi(A_{i2}, A_{i1}) \leq 1$, and $0 \leq \pi(A_{i1}, A_{i2}) + \pi(A_{i2}, A_{i1}) \leq 1$.

According to Brans and De Smet (2016), the closer $\pi(A_{i1}, A_{i2})$ is to 0, the weaker the preference is for A_{i1} over A_{i2}; and the closer it is to 1, the stronger the preference is.

The positive flow, representing how an alternative is preferred to the other $m - 1$ alternatives, is

$$\varnothing^+(A_{i0}) = \frac{1}{m-1} \sum_{\substack{1 \le i \le m \\ i \ne 0}} \pi(A_{i0}, A_i). \tag{4.8}$$

Similarly, the negative flow illustrates how an alternative is outranked by the remaining alternatives:

$$\varnothing^-(A_{i0}) = \frac{1}{m-1} \sum_{\substack{1 \le i \le m \\ i \ne 0}} \pi(A_i, A_{i0}). \tag{4.9}$$

PROMETHEE, especially PROMETHEE I, distinguishes three states:

- A_{i1} is preferred to A_{i2}, denoted $A_{i1} P^I A_{i2}$, if

$$\begin{cases} \varnothing^+(A_{i1}) > \varnothing^+(A_{i2}) \;\; and \;\; \varnothing^-(A_{i1}) < \varnothing^-(A_{i2}), \;\; or \\ \varnothing^+(A_{i1}) = \varnothing^+(A_{i2}) \;\; and \;\; \varnothing^-(A_{i1}) < \varnothing^-(A_{i2}), \;\; or. \\ \varnothing^+(A_{i1}) > \varnothing^+(A_{i2}) \;\; and \;\; \varnothing^-(A_{i1}) = \varnothing^-(A_{i2}); \end{cases} \tag{4.10}$$

- A_{i1} and A_{i2} are indifferent, denoted $A_{i1} I^I A_{i2}$, if

$$\varnothing^+(A_{i1}) = \varnothing^+(A_{i2}) \;\; and \;\; \varnothing^-(A_{i1}) = \varnothing^-(A_{i2}). \tag{4.11}$$

- A_{i1} and A_{i2} are incomparable, denoted $A_{i1} R^I A_{i2}$, if

$$\begin{cases} \varnothing^+(A_{i1}) > \varnothing^+(A_{i2}) \;\; and \;\; \varnothing^-(A_{i1}) > \varnothing^-(A_{i2}), \;\; or \\ \varnothing^+(A_{i1}) < \varnothing^+(A_{i2}) \;\; and \;\; \varnothing^-(A_{i1}) < \varnothing^-(A_{i2}); \end{cases} \tag{4.12}$$

PROMETHEE II aims for a complete ranking; thus it computes the net outranking flow as

$$\varnothing(A_{i0}) = \varnothing^+(A_{i0}) - \varnothing^-(A_{i0}). \tag{4.13}$$

It should be noted that all alternatives are comparable when using PROMETHEE II. Thus the possible cases are the following:
A_{i1} is preferred to A_{i2}, denoted $A_{i1} P^{II} A_{i2}$, if $\varnothing(A_{i1}) > \varnothing(A_{i2})$.
A_{i1} and A_{i2} are indifferent, denoted $A_{i1} I^{II} A_{i2}$, if $\varnothing(A_{i1}) = \varnothing(A_{i2})$.

To verify the accuracy of the calculations, the following must be true:

$$-1 \le \varnothing(A_i) \le 1 \;\; and \;\; \sum_i \varnothing(A_i) = 1.$$

Frequently Used MCDM Methods

TABLE 4.9
PROMETHEE Example

Criterion Type Parameter Values Nature		C_1 U-shape $q_1 = 10$ Min	C_2 V-shape $p_2 = 30$ Max	C_3 Linear $p_3 = 50, q_3 = 450$ Min	C_4 Level $p_4 = 10, q_5 = 50$ Min	C_5 Usual – Min	C_6 Gaussian $s_6 = 5$ Max
Alternative	A_1	80	90	600	54	8	5
	A_2	65	58	200	97	1	1
	A_3	83	60	400	72	4	7
	A_4	40	80	1,000	75	7	10
	A_5	52	72	600	20	3	8
	A_6	94	96	700	36	5	6

In an illustrative example (Alinezhad & Khalili, 2019), six alternatives are to be evaluated in accordance with six criteria. Table 4.9 details the evaluations in addition to criterion weights and types.

By computing the distance between each pair of evaluations, we obtain Table 4.10.

Then based on distance values in addition to the nature of the criteria (Equations 4.4 and 4.5), the type, and the relative parameters (Table 4.8), the preferences are given in Table 4.11.

Supposing that all criteria have the same weight (1/6), and using Equations 4.6 and 4.7, the preference index is computed as in Table 4.12.

The positive (Equation 4.8), negative (Equation 4.9), and net flows (Equation 4.13) are given in Table 4.13.

Based on Equations 4.10–4.12, the preferences in PROMETHEE I are detailed in Table 4.14.

Thus, the final ranking using PROMETHEE II is

$$A_5 > A_2 > A_4 > A_6 > A_3 > A_1.$$

Links between PROMETHEE and uncertainty have been presented (Giannopoulos & Founti, 2010; Hyde, Maier, & Colby, 2003), as have links with fuzzy logic (Goumas & Lygerou, 2000). Reviews of PROMETHEE have been published by Behzadian et al. (2010) and Brans and De Smet (2016).

VIKOR

VIKOR (Duckstein & Opricovic, 1980; from the Serbian Vise Kriterijumska Optimizacija i Kompromisno Resenje, which translates to Multicriteria

TABLE 4.10
Distance between Each Pair of Evaluations

Distance	Criterion					
	C_1	C_2	C_3	C_4	C_5	C_6
$d(A_1, A_2)$	15	32	400	−43	7	4
$d(A_1, A_3)$	−3	30	200	−18	4	−2
$d(A_1, A_4)$	40	10	−400	−21	1	−5
$d(A_1, A_5)$	28	18	0	34	5	−3
$d(A_1, A_6)$	15	32	400	−43	7	4
$d(A_2, A_1)$	−15	−32	−400	43	−7	−4
$d(A_2, A_3)$	−18	−2	−200	25	−3	−6
$d(A_2, A_4)$	25	−22	−800	22	−6	−9
$d(A_2, A_5)$	13	−14	−400	77	−2	−7
$d(A_2, A_6)$	−15	−32	−400	43	−7	−4
$d(A_3, A_1)$	3	−30	−200	18	−4	2
$d(A_3, A_2)$	18	2	200	−25	3	6
$d(A_3, A_4)$	43	−20	−600	−3	−3	−3
$d(A_3, A_5)$	31	−12	−200	52	1	−1
$d(A_3, A_6)$	−11	−36	−300	36	−1	1
$d(A_4, A_1)$	−40	−10	400	21	−1	5
$d(A_4, A_2)$	−25	22	800	−22	6	9
$d(A_4, A_3)$	−43	20	600	3	3	3
$d(A_4, A_5)$	−12	8	400	55	4	2
$d(A_4, A_6)$	−54	−16	300	39	2	4
$d(A_5, A_1)$	−28	−18	0	−34	−5	3
$d(A_5, A_2)$	−13	14	400	−77	2	7
$d(A_5, A_3)$	−31	12	200	−52	−1	1
$d(A_5, A_4)$	12	−8	−400	−55	−4	−2
$d(A_5, A_6)$	−42	−24	−100	−16	−2	2
$d(A_6, A_1)$	14	6	100	−18	−3	1
$d(A_6, A_2)$	29	38	500	−61	4	5
$d(A_6, A_3)$	11	36	300	−36	1	−1
$d(A_6, A_4)$	54	16	−300	−39	−2	−4
$d(A_6, A_5)$	42	24	100	16	2	−2

Optimization and Compromise Solution) happens to be the closest to the philosophy of TOPSIS (Opricovic & Tzeng, 2004), because it aims to choose the closest alternative to the ideal solution. The closeness measurement is based on the L_p metric (Alinezhad & Khalili, 2019):

TABLE 4.11
Preference-Function Values

Alternative	A_1	A_2	A_3	A_4	A_5	A_6
A_1	–	1.176	1.5	1.608	0.6	1.11
A_2	2.772	–	2.388	1.998	1.776	3
A_3	1.416	1.08	–	1.998	0.336	2.574
A_4	2.394	3.03	1.83	–	1.338	1.272
A_5	2.64	3.09	2.922	2.38	–	2.688
A_6	1.716	2.394	1.5	2.592	0.798	–

TABLE 4.12
Preference Index

Alternative	A_1	A_2	A_3	A_4	A_5	A_6
A_1	–	0.296	0.25	0.268	0.1	1.185
A_2	0.462	–	0.389	0.333	0.296	0.5
A_3	0.236	0.18	–	0.333	0.056	0.429
A_4	0.399	0.505	0.305	–	0.223	0.212
A_5	0.444	0.515	0.487	0.38	–	0.448
A_6	0.286	0.399	0.25	0.432	0.133	–

TABLE 4.13
Positive, Negative, and Net Flows

Alternative	Positive Flow $\varnothing^+$	Negative Flow $\varnothing^-$	Net Flow $\varnothing$
A_1	0.22	0.365	−0.145
A_2	0.396	0.379	0.017
A_3	0.247	0.336	−0.089
A_4	0.329	0.349	−0.02
A_5	0.455	0.162	0.293
A_6	0.3	0.355	−0.055

TABLE 4.14
Preferences in PROMETHEE I

Alternative	A_1	A_2	A_3	A_4	A_5	A_6
A_1	I^I	–	–	–	–	–
A_2	–	I^I	–	–	–	–
A_3	P^I	–	I^I	–	–	–
A_4	P^I	–	–	I^I	–	P^I
A_5	P^I	P^I	P^I	P^I	I^I	P^I
A_6	P^I	–	–	–	–	I^I

$$L_{p,i} = \left\{ \sum_{j=1}^{m} \left[\frac{w_j(f_j^* - f_{ij})}{(f_j^* - f_j^-)} \right]^p \right\}^{1/p}, \quad (4.14)$$

such that $1 \leq p \leq \infty$ is the parameter identifying the L_p metric,
f_{ij} is the value of the ith alternative in accordance with the jth criterion,
f_j^* is the best f_{ij}, and f_j^- is the worst f_{ij}.
For benefit (positive) criteria,

$$\begin{cases} f_j^* = \max_i f_{ij} \\ f_j^- = \min_i f_{ij} \end{cases} \quad (4.15)$$

For cost (negative) criteria,

$$\begin{cases} f_j^* = \min_i f_{ij} \\ f_j^- = \max_i f_{ij} \end{cases} \quad (4.16)$$

The $L_{1,i}$ metric is denoted S_i in the method, such that

$$S_i = \sum_{j=1}^{m} \frac{w_j(f_j^* - f_{ij})}{(f_j^* - f_j^-)}, \quad (4.17)$$

whereas the $L_{\infty,i}$ metric is denoted R_i:

$$R_i = \max_j \left[\frac{w_j(f_j^* - f_{ij})}{(f_j^* - f_j^-)} \right]. \quad (4.18)$$

Thus, the VIKOR index is computed for each alternative as follows:

$$Q_i = v\frac{S_i - S^*}{S^- - S^*} + (1 - v)\frac{R_i - R^*}{R^- - R^*}, \qquad (4.19)$$

where $S^* = \min_i S_i$, $S^- = \max_i S_i$, $R^* = \min_i R_i$, $R^- = \max_i R_i$,

v is the weight of the strategy of maximum group utility, and $1 - v$ is the weight of individual regret.

Then alternatives are ranked by the values of R, S, and Q in decreasing order, resulting in three ranking lists.

The chosen compromise solution $A^{(1)}$ corresponds to the minimum value of Q if it satisfies (Cristobal, 2012):

- $Q(A^{(2)}) - Q(A^{(1)}) \geq 1/(m - 1)$, with $A^{(2)}$ being the second alternative in accordance with Q.
- must also be ranked first in accordance with S and/or R.

If the last condition is not fulfilled, then both $A^{(1)}$ and $A^{(2)}$ are included.

Otherwise, include $A^{(1)}$, $A^{(2)}$, and $A^{(M)}$ such that $Q(A^{(M)}) - Q(A^{(1)}) \geq 1/(m - 1)$.

In an example treated by Opricovic and Tzeng (2004), three alternatives A_1, A_2, and A_3 are evaluated in accordance with two criteria C_1 and C_2 (Table 4.15), with C_1 being a cost criterion and C_2 a benefit criterion.

Applying Equation 4.16 for cost criterion C_1, $f_1^* = \min(1, 2, 5) = 1$ and $f_1^- = \max(1, 2, 5) = 5$.

Applying Equation 4.15 for benefit criterion C_2,

$$f_2^- = \min(3\ 000,\ 3\ 750,\ 4\ 500) = 3\ 000 \text{ and}$$

$$f_2^* = \max(3\ 000,\ 3\ 750,\ 4\ 500) = 4\ 500.$$

TABLE 4.15
VIKOR Example

	C_1	C_2	S_i	R_i	Q_i
A_1	1	3,000	0.5	0.5	1
A_2	2	3,750	3/8	0.25	0
A_3	5	4,500	0.5	0.5	1
Best (f_j^*, S^*, R^*)	1	4,500	3/8	0.25	–
Worst (f_j^-, S^-, R^-)	5	3,000	0.5	0.5	–

Supposing that the criteria have the same weight, $w_1 = w_2 = 0.5$, S_i and R_i values are computed using Equations 4.17 and 4.18, respectively. Additionally computing $S*$, S^-, $R*$ and R^-, values of Q_i are deduced using Equation 4.19; the final ranking is

$$A_2 > A_3 = A_1.$$

While the original method is content with these calculations, recent extensions including a stability test have been introduced (Opricovic, 2008; Opricovic & Tzeng, 2007). Recent reviews discussing VIKOR extensions and applications have been presented by Gul et al. (2016) and Mardani et al. (2016).

Other methods have been discussed by Alinezhad and Khalili (2019), Cristobal (2012), Greco et al. (2016), and Papathanasiou et al. (2016). Further comparisons can be found between TOPSIS and SAW (the weighted sum) (Sahu et al., 2019; Sunarti et al., 2018), AHP (Widianta et al., 2018), PROMETHEE (Kamble, Vadirajacharya, & Patil, 2019; Pangaribuan & Beniyanto, 2018), VIKOR (Ameri, Pourghasemi, & Cerda, 2018), and other methods (Lin, Chiang, & Chang, 2007).

CONCLUSION

Based on the fact that no single method can outperform all others in all aspects, this chapter insists that there is no absolute best method (TOPSIS included), but rather that certain methods are better adapted to specific situations. Thus, after reviewing the principal common steps in a decision-making process, the chapter presents through illustrative examples the most-used methods, discussing their pros and cons.

REFERENCES

A. Baky, Ibrahim, & Mahmoud A. Abo-Sinna. (2013). TOPSIS for Bi-Level MODM Problems. *Applied Mathematical Modelling*, *37*(3), 1004–1015. https://doi.org/10.1016/j.apm.2012.03.002.

Aguarón, Juan, María Teresa Escobar, José María Moreno-Jiménez, & Alberto Turón. (2019). AHP-Group Decision Making Based on Consistency. *Mathematics*, *7*(3), 242. https://doi.org/10.3390/math7030242.

Alinezhad, Alireza, & Javad Khalili. (2019). *New Methods and Applications in Multiple Attribute Decision Making (MADM)*. International Series in Operations Research & Management Science 277. Springer International Publishing. https://doi.org/10.1007/978-3-030-15009-9.

Ameri, Alireza Arab, Hamid Reza Pourghasemi, & Artemi Cerda. (2018). Erodibility Prioritization of Sub-Watersheds Using Morphometric Parameters Analysis and Its Mapping: A Comparison among TOPSIS, VIKOR, SAW, and CF Multi-Criteria Decision Making Models. *Science of The Total Environment*, *613–614*(February), 1385–1400. https://doi.org/10.1016/j.scitotenv.2017.09.210.

Asadabadi, Mehdi Rajabi, Elizabeth Chang, & Morteza Saberi. (2019). Are MCDM Methods Useful? A Critical Review of Analytic Hierarchy Process (AHP) and Analytic Network Process (ANP). Edited by Zude Zhou. *Cogent Engineering*, *6*(1), 1623153. https://doi.org/10.1080/23311916.2019.1623153.

Behzadian, Majid, R. B. Kazemzadeh, A. Albadvi, & M. Aghdasi. (2010). PROMETHEE: A Comprehensive Literature Review on Methodologies and Applications. *European Journal of Operational Research*, *200*(1), 198–215. https://doi.org/10.1016/j.ejor.2009.01.021.

Belton, Valerie, & Theo Stewart. (2002). *Multiple Criteria Decision Analysis: An Integrated Approach*. New York: Springer US. https://doi.org/10.1007/978-1-4615-1495-4.

Bhushan, Navneet, & Kanwal Rai. (2004). *Strategic Decision Making: Applying the Analytic Hierarchy Process*. Decision Engineering. London: Springer-Verlag. https://doi.org/10.1007/b97668.

Brans, J. Pierre, & Bertrand Mareschal. (1992). Promethee V: Mcdm Problems With Segmentation Constraints. *INFOR: Information Systems and Operational Research*, *30*(2), 85–96. https://doi.org/10.1080/03155986.1992.11732186.

Brans, Jean-Pierre. (1982). L'ingénieurie de La Décision-Elaboration d'instruments d'aide à La Décision. La Méthode Prométhée–Dans Nadeau R. et Landry M. *L'aide à La Décision: Nature, Intruments et Perspectives d'avenir. Presses* (pp. 182–213). Quebec, Canada.

Brans, Jean-Pierre, & Yves De Smet. (2016). PROMETHEE Methods. In Salvatore Greco, Matthias Ehrgott, & José Rui Figueira edited by, *Multiple Criteria Decision Analysis: State of the Art Surveys* (pp. 187–219). International Series in Operations Research & Management Science. New York, NY: Springer. https://doi.org/10.1007/978-1-4939-3094-4_6.

Brans, Jean-Pierre, & Bertrand Mareschal. (1995). The PROMETHEE VI Procedure: How to Differentiate Hard from Soft Multicriteria Problems. *Journal of Decision Systems*, *4*(3), 213–223. https://doi.org/10.1080/12460125.1995.10511652.

Brans, Jean-Pierre, & Bertrand Mareschal. (2001). *Prométhée - Gaia: Une méthodologie d'aide à la décision en présence de critères multiples*. Bruxelles: Ellipses Marketing.

Brans, Jean-Pierre, & Stephen James Marshall. (1994). The PROMETHEE-GAIA Decision Support System for Multicriteria Investigations. *Investigation Operativa*, *4*(2), 107–117.

Brunelli, Matteo. (2015). *Introduction to the Analytic Hierarchy Process*. Springer Briefs in Operations Research. Springer International Publishing. https://doi.org/10.1007/978-3-319-12502-2.

Chen, Ting-Yu. (2014). A PROMETHEE-Based Outranking Method for Multiple Criteria Decision Analysis with Interval Type-2 Fuzzy Sets. *Soft Computing*, *18*(5), 923–940. https://doi.org/10.1007/s00500-013-1109-4.

Clímaco, João Carlos Namorado, João Carlos Soares de Mello, & Lidia Angulo Meza. (2008). Performance Measurement: From DEA to MOLP. *Encyclopedia of Decision Making and Decision Support Technologies*, 709–715. https://doi.org/10.4018/978-1-59904-843-7.ch079.

Cristobal, Jose Ramon San. (2012). *Multi Criteria Analysis in the Renewable Energy Industry*. 1st ed. Green Energy and Technology. London: Springer-Verlag. https://doi.org/10.1007/978-1-4471-2346-0.

Dey, Partha Pratim, Surapati Pramanik, & Bibhas C. Giri. (2014). TOPSIS Approach to Linear Fractional Bi-Level MODM Problem Based on Fuzzy Goal Programming. *Journal of Industrial Engineering International*, *10*(4), 173–184. https://doi.org/10.1007/s40092-014-0073-7.

Dong, Yucheng, Zhi-Ping Fan, & Shui Yu. (2015). Consensus Building in a Local Context for the AHP-GDM With the Individual Numerical Scale and Prioritization Method. *IEEE Transactions on Fuzzy Systems*, *23*(2), 354–368. https://doi.org/10.1109/TFUZZ.2014.2312974.

Duckstein, Lucien, & Serafim Opricovic. (1980). Multiobjective Optimization in River Basin Development. *Water Resources Research*, *16*(1), 14–20. https://doi.org/10.1029/WR016i001p00014.

Dymova, Ludmila, Pavel Sevastjanov, & Anna Tikhonenko. (2016). The TOPSIS Method in the Interval Type-2 Fuzzy Setting. In Roman Wyrzykowski, Ewa Deelman, Jack Dongarra, Konrad Karczewski, Jacek Kitowski, & Kazimierz Wiatr edited by, *Parallel Processing and Applied Mathematics* (pp. 445–454). Lecture Notes in Computer Science. Switzerland: Springer International Publishing.

El Alaoui, Mohamed, & Hussain Ben-azza. (2017a). Aggregation of Performance Indicators for Supply Chain and Fuzzy Logic Extensions Applied to Green Supply Chain. In *2017 International Colloquium on Logistics and Supply Chain Management (LOGISTIQUA)*, 36–41. https://doi.org/10.1109/LOGISTIQUA.2017.7962870.

El Alaoui, Mohamed, & Hussain Ben-azza. (2017b). Generalization of the Weighted Product Aggregation Applied to Data Fusion of Intuitionistic Fuzzy Quantities. *2017 Intelligent Systems and Computer Vision (ISCV)*, 1–6. https://doi.org/10.1109/ISACV.2017.8054908.

El Alaoui, Mohamed, Hussain Ben-azza, & Khalid El Yassini. (2019). Fuzzy TOPSIS with Coherent Measure: Applied to a Closed Loop Agriculture Supply Chain. In Mostafa Ezziyyani edited by, *Advanced Intelligent Systems for Sustainable Development (AI2SD'2018)* (pp. 106–117). Advances in Intelligent Systems and Computing. Switzerland: Springer International Publishing. https://doi.org/10.1007/978-3-030-11878-5_12.

El Alaoui, Mohamed, Khalid El Yassini, & Hussain Ben-azza. (2019). Type 2 Fuzzy TOPSIS for Agriculture MCDM Problems. *International Journal of Sustainable Agricultural Management and Informatics*, 5(2/3), 112–130. https://doi.org/10.1504/IJSAMI.2019.101672.

Emrouznejad, Ali, William Ho, & William Ho. (2017). *Fuzzy Analytic Hierarchy Process*. Chapman and Hall/CRC. https://doi.org/10.1201/9781315369884.

Emrouznejad, Ali, & Marianna Marra. (2017). The State of the Art Development of AHP (1979–2017): A Literature Review with a Social Network Analysis. *International Journal of Production Research*, 55(22), 6653–6675. https://doi.org/10.1080/00207543.2017.1334976.

Figueira, José J. R., Yves De Smet, & Jean Pierre Brans. (2004). MCDA methods for sorting and clustering problems: Promethee TRI and Promethee CLUSTER. Université Libre de Bruxelles. Service de Mathématiques et de gestion. http://hdl.handle.net/2013/.

Figueira, José Rui, Salvatore Greco, Bernard Roy, & Roman Słowiński. (2013). An Overview of ELECTRE Methods and Their Recent Extensions. *Journal of Multi-Criteria Decision Analysis*, 20(1–2), 61–85. https://doi.org/10.1002/mcda.1482.

Figueira, José Rui, Vincent Mousseau, & Bernard Roy. (2016). ELECTRE Methods. In Salvatore Greco, Matthias Ehrgott, & José Rui Figueira edited by, *Multiple Criteria Decision Analysis: State of the Art Surveys* (pp. 155–185). International Series in Operations Research & Management Science. New York, NY: Springer. https://doi.org/10.1007/978-1-4939-3094-4_5.

Giannopoulos, D., & M. Founti. (2010). A Fuzzy Approach to Incorporate Uncertainty in the PROMETHEE Multicriteria Method. *International Journal of Multicriteria Decision Making*, 1(1), 80–102. https://doi.org/10.1504/IJMCDM.2010.033688.

Golden, Bruce L., Edward A. Wasil, & Patrick T. Harker. (Eds.), (1989). *The Analytic Hierarchy Process: Applications and Studies*. Berlin, Heidelberg: Springer-Verlag. https://doi.org/10.1007/978-3-642-50244-6.

Goumas, M., & V. Lygerou. (2000). An Extension of the PROMETHEE Method for Decision Making in Fuzzy Environment: Ranking of Alternative Energy Exploitation Projects. *European Journal of Operational Research*, 123(3), 606–613. https://doi.org/10.1016/S0377-2217(99)00093-4.

Govindan, Kannan, & Martin Brandt Jepsen. (2016). ELECTRE: A Comprehensive Literature Review on Methodologies and Applications. *European Journal of Operational Research*, *250*(1), 1–29. https://doi.org/10.1016/j.ejor.2015.07.019.

Greco, Salvatore, Matthias Ehrgott, & José Rui Figueira. (Eds.), (2016). *Multiple Criteria Decision Analysis: State of the Art Surveys* (*Vol. 233*). International Series in Operations Research & Management Science. New York, NY: Springer New York. https://doi.org/10.1007/978-1-4939-3094-4.

Guitouni, Adel, & Jean-Marc Martel. (1998). Tentative Guidelines to Help Choosing an Appropriate MCDA Method. *European Journal of Operational Research*, *109*(2), 501–521. https://doi.org/10.1016/S0377-2217(98)00073-3.

Gul, Muhammet, Erkan Celik, Nezir Aydin, Alev Taskin Gumus, & Ali Fuat Guneri. (2016). A State of the Art Literature Review of VIKOR and Its Fuzzy Extensions on Applications. *Applied Soft Computing*, *46*(September), 60–89. https://doi.org/10.1016/j.asoc.2016.04.040.

Hamadameen, Abdulqader Othman, & Nasruddin Hassan. (2018). A Compromise Solution For The Fully Fuzzy Multiobjective Linear Programming Problems. *IEEE Access*, *6*(August), 43696–43711. https://doi.org/10.1109/ACCESS.2018.2863566.

Hyde, Kylie, Holger R. Maier, & Christopher Colby. (2003). Incorporating Uncertainty in the PROMETHEE MCDA Method. *Journal of Multi-Criteria Decision Analysis*, *12*(4–5), 245–259. https://doi.org/10.1002/mcda.361.

Kadziński, Miłosz, Salvatore Greco, & Roman Słowiński. (2012). Extreme Ranking Analysis in Robust Ordinal Regression. *Omega*, *40*(4), 488–501. https://doi.org/10.1016/j.omega.2011.09.003.

Kamble, S. G., K. Vadirajacharya, & U. V. Patil. (2019). Comparison of Multiple Attribute Decision-Making Methods—TOPSIS and PROMETHEE for Distribution Systems. In Brijesh Iyer, S. L. Nalbalwar, & Nagendra Prasad Pathak edited by, *Computing, Communication and Signal Processing* (pp. 669–680). Advances in Intelligent Systems and Computing. Singapore: Springer. https://doi.org/10.1007/978-981-13-1513-8_68.

Kou, Gang, Daji Ergu, Yi Peng, & Yong Shi. (2013). *Data Processing for the AHP/ANP*. Quantitative Management. Berlin, Heidelberg: Springer-Verlag. https://doi.org/10.1007/978-3-642-29213-2.

Kubler, Sylvain, Jérémy Robert, William Derigent, Alexandre Voisin, & Yves Le Traon. (2016). A State-of the-Art Survey & Testbed of Fuzzy AHP (FAHP) Applications. *Expert Systems with Applications*, *65*(December), 398–422. https://doi.org/10.1016/j.eswa.2016.08.064.

Kunasekaran, Viveksarathi, & Kannan Krishnamoorthy. (2014). Multi Criteria Decision Making to Select the Best Method for the Preparation of Solid Lipid Nanoparticles of Rasagiline Mesylate Using Analytic Hierarchy Process. *Journal of Advanced Pharmaceutical Technology & Research*, *5*(3), 115–121. https://doi.org/10.4103/2231-4040.137410.

Kurka, Thomas, & David Blackwood. (2013). Selection of MCA Methods to Support Decision Making for Renewable Energy Developments. *Renewable and Sustainable Energy Reviews*, *27*(November), 225–233. https://doi.org/10.1016/j.rser.2013.07.001.

Lai, Young-Jou, Ting-Yun Liu, & Ching-Lai Hwang. (1994). TOPSIS for MODM. *European Journal of Operational Research*, Facility Location Models for Distribution Planning, *76*(3), 486–500. https://doi.org/10.1016/0377-2217(94)90282-8.

Li, Lushu, & K. K. Lai. (2000). A Fuzzy Approach to the Multiobjective Transportation Problem. *Computers & Operations Research*, *27*(1), 43–57. https://doi.org/10.1016/S0305-0548(99)00007-6.

Lin, Jyh-Jiuan, Miao-Chen Chiang, & Ching-Hui Chang. (2007). A Comparison of Usual Indices and Extended TOPSIS Methods in Mutual Funds' Performance Evaluation.

Journal of Statistics and Management Systems, *10*(6), 869–883. https://doi.org/10.1080/09720510.2007.10701289.

Mardani, Abbas, Edmundas Kazimieras Zavadskas, Kannan Govindan, Aslan Amat Senin, & Ahmad Jusoh. (2016). VIKOR Technique: A Systematic Review of the State of the Art Literature on Methodologies and Applications. *Sustainability*, *8*(1), 37. https://doi.org/10.3390/su8010037.

Mareschal, Bertrand, Jean Pierre Brans, & Cathy Macharis. (1998). The GDSS PROMETHEE Procedure: A PROMETHEE-GAIA Based Procedure for Group Decision Support. *Journal of Decision Systems*, *7*, 283–307.

Mareschal, Bertrand, & Jean-Pierre Brans. (1988). Geometrical Representations for MCDA. *European Journal of Operational Research*, *34*(1), 69–77. https://doi.org/10.1016/0377-2217(88)90456-0.

Mareschal, Bertrand, Jean-Pierre Brans, & Philippe Vincke. (1984). PROMETHEE: A New Family of Outranking Methods in Multicriteria Analysis. *Operational Research - ORIJ*, *84*(January), 477–490.

Maystre, Lucien Yves, Jacques Pictet, & Jean Simos. (1994). *Méthodes multicritères ELECTRE: description, conseils pratiques et cas d'application à la gestion environnementale*. Lausanne: PPUR presses polytechniques.

Mendoza, G. A., & H. Martins. (2006). Multi-Criteria Decision Analysis in Natural Resource Management: A Critical Review of Methods and New Modelling Paradigms. *Forest Ecology and Management*, *230*(1), 1–22. https://doi.org/10.1016/j.foreco.2006.03.023.

Mu, Enrique, & Milagros Pereyra-Rojas. (2017). AHP Models with Sub-Criteria. In Enrique Mu and Milagros Pereyra-Rojas edited by, *Practical Decision Making: An Introduction to the Analytic Hierarchy Process (AHP) Using Super Decisions V2* (pp. 45–56). Springer Briefs in Operations Research. Cham: Springer International Publishing. https://doi.org/10.1007/978-3-319-33861-3_4.

Mu, Enrique, & Milagros Pereyra-Rojas. (2018). *Practical Decision Making Using Super Decisions v3: An Introduction to the Analytic Hierarchy Process*. Springer Briefs in Operations Research. Springer International Publishing. https://doi.org/10.1007/978-3-319-68369-0.

Mukherjee, Krishnendu. (2014). Analytic Hierarchy Process and Technique for Order Preference by Similarity to Ideal Solution: A Bibliometric Analysis 'from' Past, Present and Future of AHP and TOPSIS. *International Journal of Intelligent Engineering Informatics*, *2*(2–3), 96–117. https://doi.org/10.1504/IJIEI.2014.066210.

Mukherjee, Krishnendu. (2017). *Supplier Selection: An MCDA-Based Approach*. Studies in Systems, Decision and Control. Springer India. https://www.springer.com/gp/book/9788132236986.

Opricovic, Serafim. (2007). A Fuzzy Compromise Solution for Multicriteria Problems. *International Journal of Uncertainty, Fuzziness and Knowledge-Based Systems*, *15*(03), 363–380. https://doi.org/10.1142/S0218488507004728.

Opricovic, Serafim. (2008). A Compromise Solution in Water Resources Planning. *Water Resources Management*, *23*(8), 1549. https://doi.org/10.1007/s11269-008-9340-y.

Opricovic, Serafim, & Gwo-Hshiung Tzeng. (2004). Compromise Solution by MCDM Methods: A Comparative Analysis of VIKOR and TOPSIS. *European Journal of Operational Research*, *156*(2), 445–455. https://doi.org/10.1016/S0377-2217(03)00020-1.

Opricovic, Serafim, & Gwo-Hshiung Tzeng. (2007). Extended VIKOR Method in Comparison with Outranking Methods. *European Journal of Operational Research*, *178*(2), 514–529. https://doi.org/10.1016/j.ejor.2006.01.020.

Özcan, Tuncay, Numan Çelebi, & Şakir Esnaf. (2011). Comparative Analysis of Multi-Criteria Decision Making Methodologies and Implementation of a Warehouse

Location Selection Problem. *Expert Systems with Applications*, *38*(8), 9773–9779. https://doi.org/10.1016/j.eswa.2011.02.022.

Pangaribuan, P., & A. Beniyanto. (2018). SAW, TOPSIS, PROMETHEE Method as a Comparison Method in Measuring Procurement of Goods and Services Auction System. *IOP Conference Series: Materials Science and Engineering*, *407*(September), 012045. https://doi.org/10.1088/1757-899X/407/1/012045.

Papathanasiou, Jason, Nikolaos Ploskas, Thomas Bournaris, & Basil Manos. (2016). A Decision Support System for Multiple Criteria Alternative Ranking Using TOPSIS and VIKOR: A Case Study on Social Sustainability in Agriculture. In *Decision Support Systems VI - Addressing Sustainability and Societal Challenges* (pp. 3–15). Lecture Notes in Business Information Processing. Springer, Cham. https://doi.org/10.1007/978-3-319-32877-5_1.

Poincaré, Henri. (2017). *La science et l'hypothèse*. Paris: Flammarion.

Rogers, Martin Gerard, Michael Bruen, & Lucien-Yves Maystre. (2000). *Electre and Decision Support: Methods and Applications in Engineering and Infrastructure Investment*. Springer US. https://doi.org/10.1007/978-1-4757-5057-7.

Romero, C., & T. Rehman. (2003). *Multiple Criteria Analysis for Agricultural Decisions, Second Edition, Volume 11*. 1 edition. Amsterdam; Boston: Elsevier Science.

Roy, B. (1968). Classement et choix en présence de points de vue multiples. *RAIRO - Operations Research - Recherche Opérationnelle*, *2*(V1), 57–75.

Roy, B.. (1977). Partial Preference Analysis and Decision Aid: The Fuzzy Outranking Relation Concept. In Bell, D., Keeney, R., and Raiffa, H. edited by, *Conflicting Objectives in Decisions* (pp. 40–75). New York: Wiley.

Roy, B., & P. Bertier. (1973). La Méthode Electre II: Une Application Au Media-Planning; the Electre II Method, an Application to Media Planning. *Operational Research: OR; Proceedings of the IFORS International Conference on Operational Research*, Operational research: OR; proceedings of the IFORS International Conference on Operational Research. - Amsterdam [u.a.]: North Holland Publ. Co., ISSN 0922-517X, ZDB-ID 148227-0. - 1973, p. 291-302,.

Roy, B., & Hugonnard J. C. (1982). Classement Des Prolongements de Lignes de Metro En Banlieue Parisienne: Présentation d'une Méthode Multicritère Originale. *Cahiers Du CERO*, *24*(2-3-4), 153–171.

Roy, Bernard, & Denis Bouyssou. (1993). *Aide multicritère à la décision: méthodes et cas*. Economica.

Roy, Bernard, & Jean Michel Skalka. (1985). *ELECTRE IS: aspects méthodologiques et guide d'utilisation*. LAMSADE, Unité associée au C.N.R.S. no 825, Université de Paris Dauphine.

Roy, Bernard, & Roman Słowiński. (2013). Questions Guiding the Choice of a Multicriteria Decision Aiding Method. *EURO Journal on Decision Processes*, *1*(1), 69–97. https://doi.org/10.1007/s40070-013-0004-7.

Russo, Rosaria de F. S. M., & Roberto Camanho. (2015). Criteria in AHP: A Systematic Review of Literature. *Procedia Computer Science*, 3rd International Conference on Information Technology and Quantitative Management, ITQM 2015, *55*(January), 1123–1132. https://doi.org/10.1016/j.procs.2015.07.081.

Saaty, Thomas L. (1977). A Scaling Method for Priorities in Hierarchical Structures. *Journal of Mathematical Psychology*, *15*(3), 234–281. https://doi.org/10.1016/0022-2496(77)90033-5.

Saaty, Thomas L. (1980). *The Analytic Hierarchy Process: Planning, Priority Setting, Resource Allocation*. New York: McGraw-Hill.

Saaty, Thomas L. (2016). The Analytic Hierarchy and Analytic Network Processes for the Measurement of Intangible Criteria and for Decision-Making. In Salvatore Greco, Matthias Ehrgott, & José Rui Figueira edited by, *Multiple Criteria Decision Analysis:*

State of the Art Surveys (pp. 363–419). International Series in Operations Research & Management Science. New York, NY: Springer. https://doi.org/10.1007/978-1-4939-3094-4_10.

Saaty, Thomas L., & Luis G. Vargas. (2006). *Decision Making with the Analytic Network Process: Economic, Political, Social and Technological Applications with Benefits, Opportunities, Costs and Risks*. International Series in Operations Research & Management Science. Springer US. https://doi.org/10.1007/0-387-33987-6.

Saaty, Thomas L., & Luis G. Vargas. (2012). *Models, Methods, Concepts & Applications of the Analytic Hierarchy Process*. 2nd ed. International Series in Operations Research & Management Science. Springer US. https://doi.org/10.1007/978-1-4614-3597-6.

Sahu, Atul Kumar, Nitin Kumar Sahu, Anoop Kumar Sahu, Mridul Singh Rajput, & Harendra Kumar Narang. (2019). T-SAW Methodology for Parametric Evaluation of Surface Integrity Aspects in AlMg3 (AA5754) Alloy: Comparison with T-TOPSIS Methodology. *Measurement*, *132*(January), 309–323. https://doi.org/10.1016/j.measurement.2018.09.037.

Schärlig, Alain. (1985). *Décider sur plusieurs critères: panorama de l'aide à la décision multicritère*. Lausanne: PPUR presses polytechniques.

Schärlig, Alain. (1996). *Pratiquer Electre et Prométhée: un complément à "Décider sur plusieurs critères."* Lausanne: PPUR presses polytechniques.

Sunarti, Jenie Sundari, Sita Anggraeni, Fernando B. Siahaan, & Jimmi. (2018). Comparison Topsis And Saw Method In The Selection Of Tourism Destination In Indonesia. In *2018 Third International Conference on Informatics and Computing (ICIC)*, 1–6. https://doi.org/10.1109/IAC.2018.8780550.

Teghem, Jacques, C. Delhaye, & Pierre L. Kunsch. (1989). An Interactive Decision Support System (IDSS) for Multicriteria Decision Aid. *Mathematical and Computer Modelling*, *12*(10), 1311–1320. https://doi.org/10.1016/0895-7177(89)90370-1.

Triantaphyllou, Evangelos. (2000). *Multi-Criteria Decision Making Methods: A Comparative Study*. Applied Optimization 44. Springer US. https://www.springer.com/us/book/9780792366072.

Tzeng, Gwo-Hshiung, & Jih-Jeng Huang. (2016). *Fuzzy Multiple Objective Decision Making*. CRC Press.

Vahdani, Behnam, S. Meysam Mousavi, H. Hashemi, M. Mousakhani, & R. Tavakkoli-Moghaddam. (2013). A New Compromise Solution Method for Fuzzy Group Decision-Making Problems with an Application to the Contractor Selection. *Engineering Applications of Artificial Intelligence*, *26*(2), 779–788. https://doi.org/10.1016/j.engappai.2012.11.005.

Vinodh, S., S. Sarangan, & S. Chandra Vinoth. (2014). Application of Fuzzy Compromise Solution Method for Fit Concept Selection. *Applied Mathematical Modelling*, *38*(3), 1052–1063. https://doi.org/10.1016/j.apm.2013.07.027.

Widianta, M. M. D., T. Rizaldi, D. P. S. Setyohadi, & H. Y. Riskiawan. (2018). Comparison of Multi-Criteria Decision Support Methods (AHP, TOPSIS, SAW & PROMENTHEE) for Employee Placement. *Journal of Physics: Conference Series*, *953*(January), 012116. https://doi.org/10.1088/1742-6596/953/1/012116.

Yoon, Kwangsun. (1987). A Reconciliation Among Discrete Compromise Solutions. *Journal of the Operational Research Society*, *38*(3), 277–286. https://doi.org/10.1057/jors.1987.44.

Zelany, Milan. (1974). A Concept of Compromise Solutions and the Method of the Displaced Ideal. *Computers & Operations Research*, *1*(3), 479–496. https://doi.org/10.1016/0305-0548(74)90064-1.

5 TOPSIS Methodology and Limits

INITIAL APPROACH

TOPSIS is a user-friendly method that require few inputs from the user (El Alaoui, Ben-azza, & El Yassini, 2019). It consists of choosing the alternative which is closest to the best solution, called the positive ideal solution, and furthest from the worst solution, called the negative ideal solution. The initial approach requires the following:

- Evaluating alternatives according to each criterion x_{ij}, where i: $(1 \leq i \leq m)$ describes alternatives and j: $(1 \leq j \leq n)$ describes criteria, producing $m \times n$ values in the evaluation matrix $(x_{ij})_{m \times n}$
- Criteria weights w_j; generally, $\sum w_j = 1$

It can be detailed as

$$X = \begin{pmatrix} & C_1 & \cdots & C_n \\ A_1 & x_{11} & \cdots & x_{1n} \\ \vdots & \vdots & \ddots & \vdots \\ A_m & x_{m1} & \cdots & x_{mn} \\ & w_1 & \cdots & w_n \end{pmatrix}.$$

The original approach contains the following six steps (Hwang & Yoon, 1981):

Step 1: Construct the normalized decision matrix.
 To ease comparisons, this step aims to produce dimensionless data as follows:

$$r_{ij} = \frac{x_{ij}}{\sqrt{\sum_{i=1}^{m} x_{ij}^2}}, \qquad (5.1)$$

where r_{ij} represents the normalized evaluation of alternative i according to criterion j.
Step 2: Construct the weighted normalized matrix.

Using criteria weights w_j, compute the weighted normalized evaluation v_{ij} as

$$v_{ij} = w_j \times r_{ij}. \qquad (5.2)$$

Step 3: Determine ideal solutions.
 The positive ideal solution is

$$A^* = \{v_1^*, \ldots, v_n^*\} = \{(\max_i v_{ij}|j \in \Omega_b), (\min_i v_{ij}|j \in \Omega_c)\}, \quad (5.3)$$

and the negative ideal solution is

$$A^- = \{v_1^-, \ldots, v_n^-\} = \{(\min_i v_{ij}|j \in \Omega_b), (\max_i v_{ij}|j \in \Omega_c)\}, \quad (5.4)$$

where Ω_b is the set of benefit criteria and Ω_c the set of cost criteria.

Step 4: Compute the distance between each alternative and the ideal solutions.
The distance to the positive ideal solution is

$$D_i^* = \sqrt{\sum_{j=1}^m (v_{ij} - v_j^*)^2}, \quad i = 1, \ldots, m, \quad (5.5)$$

and the distance to the negative ideal solution is

$$D_i^- = \sqrt{\sum_{j=1}^m (v_{ij} - v_j^-)^2}, \quad i = 1, \ldots, m. \quad (5.6)$$

Step 5: Calculate the relative closeness to the ideal solutions:

$$CC_i = \frac{D_i^-}{D_i^- + D_i^*}, \quad i = 1, \ldots, m. \quad (5.7)$$

Step 6: Rank according to CC_i values.

The closer CC_i is to 1, the higher the corresponding alternative should be ranked. ☐

In this illustrative example (Table 5.1) proposed by Hwang and Yoon (1981), four fighter jets—alternatives (A_1, A_2, A_3, and A_4)—are evaluated according to six criteria (C_1: maximum speed; C_2: ferry range; C_3: maximum payload; C_4: cost; C_5: reliability; and C_6: maneuverability).

Step 1: Construct the normalized decision matrix.
Using Equation 5.1, the normalized evaluations of the alternatives are presented in Table 5.2.

Step 2: Construct the weighted normalized matrix.
Using Equation 5.2, the weighted normalized decisions are detailed in Table 5.3.

Step 3: determine ideal solutions.
Using Equations 5.3 and 5.4 and Table 5.1 to determine the nature of the criteria, the ideal solutions are given in Table 5.4.

Step 4: Compute the distance between each alternative and the ideal solutions.
Using Equations 5.5 and 5.6, the distances D_i^* and D_i^- to the ideal solutions are presented in Table 5.5.

Step 5: Calculate the relative closeness to the ideal solutions.
Using Equation 5.7, the closeness coefficients CC_i are also presented in Table 5.5.

Step 6: Rank according to CC_i values.
Table 5.5 shows the CC_i values and the final ranking.

TOPSIS Methodology and Limits

TABLE 5.1
Original TOPSIS Example

Criteria Nature		C_1 Max	C_2 Max	C_3 Max	C_4 Min	C_5 Max	C_6 Max
Weight		0.2	0.1	0.1	0.1	0.2	0.3
Alternative	A_1	2	1,500	20,000	5.5	5	9
	A_2	2.5	2,700	18,000	6.5	3	5
	A_3	1.8	2,000	21,000	4.5	7	7
	A_4	2.2	1,800	20,000	5	5	5

TABLE 5.2
Normalized Evaluations of Alternatives

Alternative	Criterion					
	C_1	C_2	C_3	C_4	C_5	C_6
A_1	0.4671	0.3662	0.5056	0.5069	0.4811	0.6708
A_2	0.5839	0.6591	0.4550	0.5990	0.2887	0.3727
A_3	0.4204	0.4882	0.5308	0.4147	0.6736	0.5217
A_4	0.5139	0.4394	0.5056	0.4608	0.4811	0.3727

TABLE 5.3
Weighted Normalized Decisions

Alternative	Criterion					
	C_1	C_2	C_3	C_4	C_5	C_6
A_1	0.0934	0.0366	0.0506	0.0507	0.0962	0.2012
A_2	0.1168	0.0659	0.0455	0.0599	0.0577	0.1118
A_3	0.0841	0.0488	0.0531	0.0415	0.1347	0.1565
A_4	0.1028	0.0439	0.0506	0.0461	0.0962	0.1118

MATLAB IMPLEMENTATION

To ease computations, a simple step-by-step MATLAB code could be written as follows:

```
% step 0: requirements
% alternative evaluation in rows according to criteria in columns
```

TABLE 5.4
Ideal Solutions

Ideal Solution	Criterion					
	C_1	C_2	C_3	C_4	C_5	C_6
Positive A^*	0.1168	0.0659	0.0531	0.0415	0.1347	0.2012
Negative A^-	0.0841	0.0366	0.0455	0.0599	0.0577	0.1118

TABLE 5.5
Distances to Ideal Solutions, Closeness Coefficients, and Final Ranking

Alternative	D_i^*	D_i^-	CC_i	Final Ranking
A_1	0.0546	0.0984	0.6433	1
A_2	0.1197	0.0439	0.2684	4
A_3	0.0580	0.0920	0.6135	2
A_4	0.1009	0.0458	0.3123	3

```
X = [2 1500 20000 5.5 5 9
2.5 2700 18000 6.5 3 5
1.8 2000 21000 4.5 7 7
2.2 1800 20000 5 5 5];
W = [0.2 0.1 0.1 0.1 0.2 0.3]; % criterion weights
BC = [1 1 1 0 1 1]; % 1: benefit criterion 0: cost criterion
% step 1: construct the normalized decision matrix
[m, n]=size(X); % m: number of alternatives n: number of
criteria
Nor=zeros(1,n); %normalizing terms 1/sqrt(x(i,j)^2)
for j=1:n;
for i=1:m
Nor(j)=Nor(j)+X(i,j)^2;
end
Nor(j)=Nor(j)^(0.5);
end
R=zeros(m,n); % normalized decision matrix
for i=1:m
for j=1:n
R(i,j)=X(i,j)/Nor(j);
end
end
% step 2: construct the weighted normalized matrix
```

```
V=zeros(m,n); % weighted normalized decision matix
for i=1:m
for j=1:n
V(i,j)=W(1,j)*R(i,j);
end
end
% step 3: determine ideal solutions
A=zeros(2,n); %A(1,:): A* PIS, A(2,:): A- NIS
for j=1:n
if BC(j)==1 % 1: benefit criterion
A(1,j)=max(V(:,j));
A(2,j)=min(V(:,j));
else % 0: cost criterion
A(1,j)=min(V(:,j));
A(2,j)=max(V(:,j));
end
end
% step 4: compute the distances between each weighted normalized alternative and the ideal solutions
D=zeros(m,2); %D(:,1): D+, D(:,2): D-
for i=1:m
for j=1:n
D(i,1)=D(i,1)+(V(i,j)-A(1,j))^2;
D(i,2)=D(i,2)+(V(i,j)-A(2,j))^2;
end
D(i,1)=D(i,1)^0.5;
D(i,2)=D(i,2)^0.5;
end
% step 5: calculate the relative closeness of each alternative
CC=zeros(m,1); % Closeness Coefficients
for i=1:m
CC(i)=D(i,2)/(D(i,1)+D(i,2));
end
% step 6: rank according to CC values
[~,idx]=ismember(CC,sort(CC,'descend')); % idx contains alternative rank
```

RANK REVERSAL

Rank reversal was, is, and probably will remain an important tool for judging the validity of a multiattribute decision-making method. Rank reversal was discussed first in the following example (Table 5.6), in which three alternatives are compared according to three criteria in pairwise comparisons, using the analytic hierarchy process (Belton & Gear, 1983).

TABLE 5.6
Comparison Matrix of Three Alternatives According to Three Criteria

Criterion	C_1			C_2			C_3		
Alternative	A_1	A_2	A_3	A_1	A_2	A_3	A_1	A_2	A_3
A_1	1	1/9	1	1	9	9	1	8/9	8
A_2	9	1	9	1/9	1	1	9/8	1	9
A_3	1	1/9	1	1/9	1	1	1/8	1/9	1

Using the analytical hierarchy process (AHP) described in Chapter 4, the final ranking is

$$A_2 \geq A_1 \geq A_3,$$

where the symbol $\geq$ means "is preferred to." Thus A_2 is preferred to A_1 and A_1 is preferred to A_3.

By introducing a fourth alternative A_4 equivalent to the second one ($A_4 \sim A_2$), as presented in Table 5.7, we can expect the ranking

$$A_4 \sim A_2 \geq A_1 \geq A_3.$$

However the ranking actually obtained is

$$A_1 \geq A_4 \sim A_2 \geq A_3.$$

It can be observed that A_2 outperformed A_1 before A_4 was introduced—denoted $A_2 \geq A_1$—but that relationship becomes $A_1 \geq A_2$ when A_4 is considered. This phenomenon, called rank reversal, occurs not only when a duplicated alternative is introduced but also when a new alternative is added.

TABLE 5.7
Comparison Matrix of Four Alternatives According to Three Criteria

Criterion	C_1				C_2				C_3			
Alternative	A_1	A_2	A_3	A_4	A_1	A_2	A_3	A_4	A_1	A_2	A_3	A_4
A_1	1	1/9	1	1/9	1	9	9	9	1	8/9	8	8/9
A_2	9	1	9	1	1/9	1	1	1	9/8	1	9	1
A_3	1	1/9	1	1/9	1/9	1	1	1	1/8	1/9	1	1/9
A_4	9	1	9	1	1/9	1	1	1	9/8	1	9	1

The literature includes discussions of the causes of rank reversal (Brans & De Smet, 2016), including the following:

- The presence of a nondiscriminating criterion (Verly & De Smet, 2013). According to Aires and Ferreira (2019), TOPSIS did not present any rank reversal due to this cause in their studied examples.
- Addition or removal of:
  o A copy of an alternative (Table 5.7)
  o A dominated alternative (Table 5.8)
  o Any given alternative
- Division into subproblems (Tables 5.9 and 5.10). When a decision problem is divided into subproblems, the previously defined preferences between alternatives change.

Here rank reversal (as in Aires et al., 2018; El Alaoui, El Yassini, & Ben-azza, 2019; Saaty & Sagir, 2009; Wang & Luo, 2009) means that the addition or deletion of an alternative may change the order of the remaining ones.

As rank reversal was only linked to AHP (Belton & Gear, 1983), Saaty (Saaty, 1987)—its founder—alongside Vargas (Saaty & Vargas, 1984), Millet (Millet & Saaty, 2000) and Forman (Forman, 1990) argued for its legitimacy for a long period. However, after scientific quarrels on the subject (Belton & Gear, 1985; Saaty, 1990; Vargas, 1994), numerous attempts have been proposed to avoid the phenomenon (Barzilai & Golany, 1994; Pérez, 1995; Schenkerman, 1994; Wang & Elhag, 2006; Wijnmalen & Wedley, 2008), and work proving that no standardization approach can eliminate it (Barzilai & Golany, 1994). Saaty acknowledged rank reversal as a weak point of the method while maintaining its usefulness in certain situations (Saaty & Sagir, 2009). According to Millet and Saaty (2000), rank reversals appear only rarely, with little impact on the final decision.

In addition to AHP (Maleki & Zahir, 2013; Ramanathan & Ramanathan, 2011) and its evolved form ANP (Kong, Wei, & Gong, 2016), other methods suffer the same problem, including PROMETHEE (Berghman et al., 2019; Mareschal, De Smet, & Nemery, 2008; Verly & De Smet, 2013), VIKOR (Ceballos, Pelta, & Lamata, 2017), ELECTRE (Wang & Triantaphyllou, 2008), and TOPSIS (Aires & Ferreira, 2019; García-Cascales & Lamata, 2012; İç, 2014; Kong, 2011; Mousavi-Nasab & Sotoudeh-Anvari, 2018; Mufazzal & Muzakkir, 2018; Senouci et al., 2016; Tang & Fang, 2018; Yang, 2020). Still, of the 130 papers analyzed by Aires et al. (2018), 76.1% link rank reversal to AHP, 16.2% to combinations of two or more methods and experiments, 3.1% to PROMETHEE, 2.3% to ELECTRE, and 2.3% to TOPSIS.

An early comparison was published by Zanakis et al. (1998). The results showed that TOPSIS results in fewer rank reversals than ELECTRE and four versions of AHP. TOPSIS was outperformed by simple additive weighting (SAW) and multiplicative exponent weighting, which both result in no rank reversal at all.

In an experiment including 10 examples (Mousavi-Nasab & Sotoudeh-Anvari, 2018), TOPSIS resulted in the worst performance, generating rank reversals in five

TABLE 5.8
Rank Reversal in TOPSIS Example

Criteria		C_1	C_2	C_3	C_4	Four Alternatives		Five Alternatives		Three Alternatives	
Weight		1/6	1/3	1/3	1/6	CC_i	Rank	CC_i	Rank	CC_i	Rank
Alternative	A_1	36	42	43	70	0.4184	3	0.4261	4	0.4319	3
	A_2	25	50	45	80	0.4858	1	0.5086	2	0.4742	2
	A_3	28	45	50	75	0.4634	2	0.5262	1	0.5007	1
	A_4	24	40	47	100	0.3915	4	0.4317	3	—	—
	A_5	30	43	40	85	—	—	0.3348	5	—	—

out of the 10 cases. The other methods studied were SAW, with no rank reversals; COmplex PRoportional ASsessment (COPRAS), with three. However, another comparison in the same study showed that TOPSIS's rank reversals were much less serious than those of VIKOR, because they did not affect the top-ranked alternatives. The study concluded a high Spearman correlation index between all pairs of TOPSIS, Complex Proportional Assessment, and SAW.

Although Wang and Luo (2009) do not provide any solutions, they show experimentally that practically all methods suffer rank reversals, including TOPSIS and SAW. In their example (Table 5.8), five alternatives (A_1, A_2, A_3, A_4 and A_5) are evaluated according to four criteria (C_1, C_2, C_3 and C_4).

When only the first four alternatives are considered, the ranking is $A_2 \geq A_3 \geq A_1 \geq A_4$. However, when the fifth alternative A_5 is added—which is the least preferred one—the ranking of the other alternatives is greatly affected: $A_3 \geq A_2 \geq A_4 \geq A_1 \geq A_5$. With just the first three alternatives, the ranking becomes $A_3 \geq A_2 \geq A_1$, which proves that rank reversal happens in TOPSIS when alternatives are added or removed.

To investigate other causes of rank reversal (Aires & Ferreira, 2019), we can split the problem with four alternatives (A_1, A_2, A_3 and A_4) in two: subproblems 1 (Table 5.9) and 2 (Table 5.10). Although the first subproblem does not present any issues, the second one illustrates a violation of transitivity: whereas A_2 dominated A_3 in the initial situation with four alternatives, that relation is inverted when only those two alternatives are considered.

We can observe that the permutations that occur include alternatives with relatively similar closeness coefficients—(A_2 and A_3) and (A_1 and A_4)—which may suggest the use of TOPSIS for each pair of alternatives. The question that remains is: considering vagueness, at what value do we need to use TOPSIS for the compared pair? This is especially true in these comparisons, since the gap between A_2 and A_3 leaves no doubt as to which is preferred, whereas the gap between A_1 and A_4 remains tiny. Moreover, using pairwise comparisons for all alternatives requires running TOPSIS $m*(m-1)/2$ times, which is cumbersome and will result in only partial ranking, as in ELECTRE I (Chapter 4). Thus, it may be reserved for doubtful cases.

Focusing on Tables 5.9–5.13, respectively treating, subproblem 1 (A_1 and A_4) in Table 5.9, subproblem 2 (A_2 and A_3) in Table 5.10, four alternatives (A_1, A_2, A_3 and A_4) in Table 5.11, five alternatives (A_1, A_2, A_3, A_4 and A_5) in Table 5.12 and three alternatives (A_1, A_2 and A_3) in Table 5.13. It is clear that the values of the weighted normalized assessments and ideal solutions change in accordance to the number and nature of the other alternatives being considered, causing rank reversal, even though neither the initial assessments nor the criterion weights have changed. This has pushed some authors (Cables, Lamata, & Verdegay, 2016; García-Cascales & Lamata, 2012) to introduce changes to the computation of ideal solutions and to the normalization procedure in TOPSIS.

IDEAL SOLUTIONS

Different authors have different understandings of ideal solutions. While some believe that the positive and negative ideal solutions are literary ideals that may not

TABLE 5.9
Subproblem 1 (A_1 and A_4)

Criteria		C_1	C_2	C_3	C_4	CC_i	Rank
Alternative	A_1	0.1387	0.2414	0.2250	0.0956	0.5087	1
	A_4	0.0925	0.2299	0.2459	0.1365	0.4913	2
Ideal Solution	Positive	0.1387	0.2414	0.2459	0.1365		
	Negative	0.0925	0.2299	0.2250	0.0956		

be reached (El Alaoui et al., 2019), others insist on the contrary that the ideal solutions used in TOPSIS as in the original approach (Hwang & Yoon, 1981) must be feasible. Thus, the positive ideal solution is composed of all the best alternatives attributes attainable and the negative ideal solution is composed of all worst alternatives attributes attainable.

A second distinction is to be made according to the interpretation of the word "solution" itself. The first group of authors considers only weighted normalized solutions; others have also introduced ideal alternatives but require a previous knowledge of the range of variation for each criterion. To ease comparisons in this section, all criteria are supposed to be benefit criteria.

Since the TOPSIS algorithm requires alternative assessments, we begin with the second category:

- García-Cascales and Lamata (2012) introduced two factious alternatives—an upper one, or utopia alternative, $A_{m+1} = \{\max_i x_{ij}, \ldots, \max_i x_{ij}\}$; and a lower one, or nadir alternative, $A_{m+2} = \{\min_i x_{ij}, \ldots, \min_i x_{ij}\}$—which can be denoted $A_{m+1} = \{Z_t, \ldots, Z_t\}$ and $A_{m+2} = \{Z_0, \ldots, Z_0\}$. Taking into consideration normalization by the max, the ideal solutions (reference points) become $A^* = \{w_1, \ldots w_n\}$ and $A^- = \{w_1 Z_0/Z_t, \ldots, w_n Z_0/Z_t\}$.
- Other authors believe that optimality is not necessary synonymous with extremal values (Cables et al., 2016). For instance, in looking for an athlete, candidates might be between 15 and 40 years old, but the optimal age might

TABLE 5.10
Subproblem 2 (A_2 and A_3)

		C_1	C_2	C_3	C_4	CC_i	Rank
Alternative	A_2	0.1110	0.2478	0.2230	0.1216	0.4795	2
	A_3	0.1243	0.2230	0.2478	0.1140	0.5205	1
Ideal Solution	Positive	0.1243	0.2478	0.2478	0.1216		
	Negative	0.1110	0.2230	0.2230	0.1140		

TOPSIS Methodology and Limits

TABLE 5.11
Weighted Normalized Assessments and Ideal Solutions (A_1, A_2, A_3 and A_4)

Weighted normalized matrix for the first 4 alternatives		Criterion			
		C_1	C_2	C_3	C_4
Alternative	A_1	0.1047	0.1576	0.1547	0.0711
	A_2	0.0727	0.1876	0.1619	0.0813
	A_3	0.0815	0.1689	0.1799	0.0762
	A_4	0.0698	0.1501	0.1691	0.1016
Ideal Solution	Positive	0.1047	0.1876	0.1799	0.1016
	Negative	0.0698	0.1501	0.1547	0.0711

TABLE 5.12
Weighted Normalized Assessments and Ideal Solutions (A_1, A_2, A_3, A_4 and A_5)

Weighted normalized matrix for the first 5 alternatives		Criterion			
		C_1	C_2	C_3	C_4
Alternative	A_1	0.0928	0.1419	0.1420	0.0631
	A_2	0.0644	0.1689	0.1486	0.0722
	A_3	0.0722	0.1520	0.1652	0.0676
	A_4	0.0619	0.1351	0.1553	0.0902
	A_5	0.0773	0.1453	0.1323	0.0767
Ideal Solution	Positive	0.0928	0.1689	0.1652	0.0902
	Negative	0.0619	0.1352	0.1321	0.0631

TABLE 5.13
Weighted Normalized Assessments and Ideal Solutions (A_1, A_2 and A_3)

Weighted normalized matrix for the first 3 alternatives		Criterion			
		C_1	C_2	C_3	C_4
Alternative	A_1	0.1154	0.1765	0.1795	0.0897
	A_2	0.0801	0.2102	0.1879	0.1025
	A_3	0.0897	0.1891	0.2088	0.0961
Ideal Solution	Positive	0.1154	0.2102	0.2088	0.1025
	Negative	0.0801	0.1765	0.1795	0.0897

be between 20 and 30. If sorted by weights, optimal candidates would all be in a certain interval. Thus, optimality could be reached—in these authors' method—in intervals. The proposed algorithm results in $A^* = \{w_1, \ldots, w_n\}$ and $A^- = \{0, \ldots, 0\}$.

In addition to these visions of weighted normalized ideal reference points, three main interpretations of ideal solutions can be found in the literature:

- In the first one, proposed in the original TOPSIS (Hwang & Yoon, 1981), the positive ideal solution is composed of all the best weighted normalized alternative attributes attainable, $A^* = \{\max_i(v_{ij}), \ldots, \max_i(v_{ij})\}$; and the negative ideal solution is composed of all the worst weighted normalized alternative attributes attainable, $A^- = \{\min_i(v_{ij}), \ldots, \min_i(v_{ij})\}$—sort of utopia and nadir points. It is evident that $0 \leq \min_i(v_{ij}) \leq \max_i(v_{ij}) \leq 1$.
- In the second, the positive and negative ideal solutions are literary ideals that may not be reached (El Alaoui e al., 2019). Thus the positive ideal solution is $A^* = \{1, 1, \ldots, 1\}$ and the negative ideal solution is $A^- = \{0, 0, \ldots, 0\}$. This is especially the case in the fuzzy context (Chen, 2000).
- A third category opts for a combination of both (Mukherjee, 2014), which produces the positive ideal solution $A^* = \{1, 1, \ldots, 1\}$ and the negative ideal solution is $A^- = \{\min_i(v_{ij}), \ldots, \min_i(v_{ij})\}$.

Other interpretations and adaptations exist in the fuzzy context.

As one of the main objectives is to reduce or eliminate rank reversals, the second interpretation, with $A^* = \{1, 1, \ldots, 1\}$ and $A^- = \{0, 0, \ldots, 0\}$, seems to be suitable, since the reference points do not depend at all on any other values. The approaches proposed by (Cables et al., 2016; García-Cascales & Lamata, 2012), despite also opting for a sort of independence between the alternatives and ideal solutions, require full knowledge of criterion variation ranges to determine the ideal alternative, which is not always possible.

In order to study the influence of the reference points, the following comparison uses the normalized weighted values from Tables 5.11–5.13, then computes the Euclidean distance to the five possible reference points:

$\{0, 0, \ldots, 0\}, \{w_1 Z_0/Z_t, \ldots, w_n Z_0/Z_t\}, \{w_1, \ldots, w_n\}, \{\min_i(v_{ij}), \ldots, \min_i(v_{ij})\},$
$\{\max_i(v_{ij}), \ldots, \max_i(v_{ij})\}$ and $\{1, 1, \ldots, 1\}$

Hence, based on the distances in Table 5.14, the closeness coefficients for each of the five possible combinations of ideal solutions in Table 5.15, along with the relative ranking.

The comparison of the rankings obtained in Table 5.15 confirms that rank reversal occurs only when the reference points depend on the alternatives compared. Even though using $\{0, \ldots, 0\}$ and $\{1, \ldots, 1\}$ (El Alaoui et al., 2019) or $\{0, \ldots, 0\}$ and $\{w_1, \ldots, w_n\}$ (Cables et al., 2016) may diverge as to which alternative should be ranked first (A_2 or A_3), having fixed ideal points totally independent from the set of alternatives prevents rank reversals.

TABLE 5.14
Distance from the Weighted Normalized Assessment to the Ideal Points
$d(v_{ij}, \cdot)$

Situation	Alternative	$\{0, 0, \ldots, 0\}$	$\{w_1 Z_0/Z_1, \ldots, w_n Z_0/Z_1\}$	$\{w_1, \ldots, w_n\}$	Reference Point $\{\min_i(v_{ij}), \ldots, \min_i(v_{ij})\}$	$\{\max_i(v_{ij}), \ldots, \max_i(v_{ij})\}$	$\{1, 1, \ldots, 1\}$
Four alternatives	A_1	0.2546	0.2123	1.2681	0.0357	0.0496	1.7574
	A_2	0.2708	0.189	1.2419	0.0396	0.042	1.7511
	A_3	0.2708	0.2088	1.255	0.0339	0.0392	1.7494
	A_4	0.2575	0.1899	1.2513	0.0337	0.0524	1.7564
Five alternatives	A_1	0.23	0.2181	1.2869	0.0332	0.0447	1.7814
	A_2	0.2449	0.1963	1.2634	0.0388	0.0374	1.7753
	A_3	0.2453	0.2139	1.275	0.0388	0.0349	1.7738
	A_4	0.2331	0.199	1.2722	0.0356	0.0469	1.7803
	A_5	0.2245	0.1994	1.2773	0.0229	0.0455	1.7854
Three alternatives	A_1	0.2911	0.279	1.2382	0.0352	0.0464	1.7212
	A_2	0.3105	0.2497	1.2075	0.0369	0.041	1.7132
	A_3	0.3109	0.2728	1.2231	0.0339	0.0338	1.7115

TABLE 5.15
Ideal Solutions Effect on Rank Reversal

Ideal solutions		$\{0, \ldots, 0\}$ and $\{1, \ldots, 1\}$ (El Alaoui, El Yassini, & Ben-azza 2019)		$\{0, \ldots, 0\}$ and $\{w_1, \ldots, w_n\}$ (Cables, Lamata, & Verdegay 2016)		$\{w_1Z_0/Z_t, \ldots, w_nZ_0/Z_t\}$ and $\{w_1, \ldots, w_n\}$ (García-Cascales & Lamata 2012)		$\{\min_i(v_{ij}), \ldots, \min_i(v_{ij})\}$ and $\{\max_i(v_{ij}), \ldots, \max_i(v_{ij})\}$ (Hwang & Yoon 1981)		$\{\min_i(v_{ij}), \ldots, \min_i(v_{ij})\}$ and $\{1, \ldots, 1\}$ (Mukherjee 2014)	
Situation	alternatives	CC_i	rank	CC_i	rank	CC_i	rank	CC_i	rank	CC_i	rank
Four alternatives	A_1	0.1265	4	0.1672	4	0.1434	1	0.4184	3	0.0199	2
	A_2	0.1339	2	0.179	1	0.1321	3	0.4858	1	0.0221	1
	A_3	0.134	1	0.1775	2	0.1426	2	0.4634	2	0.019	3
	A_4	0.1279	3	0.1707	3	0.1318	4	0.3915	4	0.0188	4
Five alternatives	A_1	0.1143	4	0.1516	4	0.1449	1	0.4261	4	0.0183	4
	A_2	0.1212	2	0.1624	1	0.1345	5	0.5086	2	0.0214	2
	A_3	0.1215	1	0.1613	2	0.1437	2	0.5262	1	0.0214	1
	A_4	0.1158	3	0.1548	3	0.1353	3	0.4317	3	0.0196	3
	A_5	0.1117	5	0.1495	5	0.1350	4	0.3348	5	0.0127	5
Three alternatives	A_1	0.1447	3	0.1904	3	0.1839	1	0.4319	3	0.0201	2
	A_2	0.1534	2	0.2045	1	0.1714	3	0.4742	2	0.0211	1
	A_3	0.1537	1	0.2027	2	0.1824	2	0.5007	1	0.0194	3

The use of classical ideal solutions $A^* = \{\max_i(v_{ij}), \ldots, \max_i(v_{ij})\}$ and $A^- = \{\min_i(v_{ij}), \ldots, \min_i(v_{ij})\}$ (Hwang & Yoon, 1981) was already described in the previous section, resulting in rank inversions between A_2 and A_3 and between A_1 and A_4.

Replacing the positive ideal solution with $A^* = \{1, 1, \ldots, 1\}$ while keeping the negative ideal solution as $A^- = \{\min_i(v_{ij}), \ldots, \min_i(v_{ij})\}$ (Mukherjee, 2014) may be more severe, since A_3 was ranked third in the problem involving four alternatives and becomes first when the irrelevant alternative A_5 (ranked last) is introduced.

Using $\{w_1 Z_0/Z_t, \ldots, w_n Z_0/Z_t\}$ and $\{w_1, \ldots, w_n\}$ (García-Cascales & Lamata, 2012) also results in rank reversals: while A_2 dominated A_4 when four alternatives were considered, the addition of A_5 inverted the preference order. Additionally, due to the reference points used, this is the only approach that ranks A_1 first, which does not seem logical, since it is dominated by A_2, A_3, and A_4 according to C_2, C_3 and C_4; all other approaches rank it fourth.

However, since the weighted normalized values for each alternative change in each situation due to the normalization used, the next section reviews some normalization procedures.

NORMALIZATION

Normalization is a fundamental step in all multicriteria decision-making methods. Falkner and Verter (1991) showed that the use of nonnormalized alternatives could produce conflicting results. According to Yu, Pan, and Wu (2009), this process of normalization should satisfy three conditions:

- The relative gap between data for the same indicator should remain constant. In a linear normalization, such as Yu et al. (2009) advocate for. If x_1 and x_2 are the raw values and r_1 and r_2 are the normalized ones, then $(r_2 - r_1)/(x_2 - x_1)$ must be constant.
- The relative gap between different indicators should remain variable. Pavličić (2001) insists on the difference between the domains of each normalization method.
- The maximum values after normalization should be equal.

Several authors have studied the effect of normalization on decision making (Chatterjee & Chakraborty, 2014; Jahan & Edwards, 2015; Milani et al., 2005), especially with TOPSIS (Acuña-Soto, Liern, & Pérez-Gladish, 2018; Chakraborty & Yeh, 2009; Chen, 2019; Sarraf, Mohaghar, & Bazargani, 2013; Vafaei, Ribeiro, & Camarinha-Matos, 2018; Zaidan & Zaidan, 2018). These studies can be divided into two main categories.

The first category states that the choice of normalization method should be adapted to the subject under study (Migilinskas & Ustinovichius, 2007). The second category advocates for a specific normalization method over others (Zavadskas, Zakarevicius, & Antucheviciene, 2006). However, the majority of methods favoring a specific normalization judge its accuracy as a function of a metric (correlation or otherwise) between the ranking produced and other rankings (Chakraborty & Yeh,

TABLE 5.16
Ten Normalization Formulas for Benefit and Cost Criteria

Formula	Benefit Criteria	Cost Criteria
Vector normalization (Jahan & Edwards, 2015)	$r_{ij} = \dfrac{x_{ij}}{\sqrt{\sum_{i=1}^{m} x_{ij}^2}}$	$r_{ij} = 1 - \dfrac{x_{ij}}{\sqrt{\sum_{i=1}^{m} x_{ij}^2}}$
Linear sum normalization (Jahan & Edwards, 2015)	$r_{ij} = \dfrac{x_{ij}}{\sum_{i=1}^{m} x_{ij}}$	$r_{ij} = \dfrac{1/x_{ij}}{\sum_{i=1}^{m} 1/x_{ij}}$
Linear max (Çelen, 2014)	$r_{ij} = \dfrac{x_{ij}}{x_j^*}$	$r_{ij} = 1 - \dfrac{x_{ij}}{x_j^*}$
Linear max-min (Jahan & Edwards, 2015)	$r_{ij} = \dfrac{x_{ij} - x_j^-}{x_j^* - x_j^-}$	$r_{ij} = \dfrac{x_j^* - x_{ij}}{x_j^* - x_j^-}$
Logarithmic (Zavadskas, Ustinovichius, & Peldschus, 2003)	$r_{ij} = \dfrac{\ln(x_{ij})}{\ln\left(\prod_{i=1}^{m} x_{ij}\right)}$	$r_{ij} = \dfrac{1 - \dfrac{\ln(x_{ij})}{\ln\left(\prod_{i=1}^{m} x_{ij}\right)}}{m - 1}$
Marković (Marković, 2010)		$r_{ij} = 1 - \dfrac{x_{ij} - x_j^-}{x_j^*}$
Tzeng and Huang (Tzeng & Huang, 2011)		$r_{ij} = \dfrac{x_j^*}{x_{ij}}$
Nonlinear normalization (Peldschus, Vaigauskas, & Zavadskas, 1983)	$r_{ij} = \left(\dfrac{x_{ij}}{x_j^*}\right)^2$	$r_{ij} = \left(\dfrac{x_j^-}{x_{ij}}\right)^2$
Lai and Hwang (Lai & Hwang, 1994)	$r_{ij} = \dfrac{x_{ij}}{x_j^* - x_j^-}$	$r_{ij} = \dfrac{x_{ij}}{x_j^- - x_j^*}$
Zavadskas and Turskis (Zavadskas & Turskis, 2008)	$r_{ij} = 1 - \left|\dfrac{x_j^* - x_{ij}}{x_j^*}\right|$	$r_{ij} = 1 - \left|\dfrac{x_j^- - x_{ij}}{x_j^-}\right|$

With $x_j^* = \max_i x_{ij}$ and $x_j^- = \max_i x_{ij}$

2009; Çelen, 2014; Vafaei et al., 2018)—which is more than questionable, since there is no absolute perfect ranking with which the comparison can be made. A simpler idea is to look for the method that prevents rank reversal, while of course providing coherent results. Furthermore, the judicious choice of a reference point alone seems insufficient to avoid the reversal of ranks. For this reason, several methods have studied a combination of reference points (ideal solutions) and normalization methods (García-Cascales & Lamata, 2012; Kong, 2011). Table 5.16 contains the ten normalization methods studied by Ploskas and Papathanasiou (2019), and Table 5.17 details the combination of the five ideal solutions and the 10 normalization procedures as applied to the example in Table 5.8.

The normalizations that do not produce any rank reversals with any combination of ideal solutions in this example are vector, linear max-min, and nonlinear normalization. However, vector normalization has elsewhere been proven to cause rank reversal (García-Cascales & Lamata, 2012).

Linear sum normalization with the ideal solutions $\{w_1 Z_0/Z_t, \ldots, w_n Z_0/Z_t\}$ and $\{w_1, \ldots, w_n\}$ (García-Cascales & Lamata, 2012) here produces a total rank reversal.

TABLE 5.17
Ideal Solutions and Normalization Effect on Rank Reversal

Ideal solutions	alternatives	$\{0,\ldots,0\}$ and $\{1,\ldots,1\}$ (El Alaoui, El Yassini, & Benazza 2019)		$\{0,\ldots,0\}$ and $\{w_1,\ldots,w_n\}$, (Cables, Lamata, & Verdegay 2016)		$\{w_1Z_0/Z_1,\ldots, w_nZ_0/Z_1\}$ and $\{w_1,\ldots, w_n\}$, (García-Cascales & Lamata 2012)		$\{\min_i(v_{ij}),\ldots,\min_i(v_{ij})\}$ and $\{\max_i(v_{ij}),\ldots,\max_i(v_{ij})\}$ (Hwang & Yoon 1981)		$\{\min_i(v_{ij}),\ldots,\min_i(v_{ij})\}$ and $\{1,\ldots,1\}$ (Mukherjee 2014)	
normalization		CC_i	rank	CC_i	rank	CC_i	rank	CC_i	rank	CC_i	rank
Vector normalization (Jahan & Edwards 2015)	A_1	0.1143	4	0.4347	4	0.2596	4	0.4261	4	0.0183	4
	A_2	0.1212	2	0.4632	2	0.2961	2	0.5086	2	0.0214	2
	A_3	0.1215	1	0.4646	1	0.2968	1	0.5262	1	0.0214	1
	A_4	0.1158	3	0.4402	3	0.2674	3	0.4317	3	0.0196	3
	A_5	0.1117	5	0.4254	5	0.2453	5	0.3348	5	0.0127	5
Linear sum normalization (Jahan & Edwards 2015)	A_1	0.0515	4	0.1958	4	0.0575	2	0.4274	4	0.0078	4
	A_2	0.0547	2	0.2084	2	0.044	4	0.5072	2	0.0091	2
	A_3	0.0549	1	0.2089	1	0.0411	5	0.5251	1	0.0091	1
	A_4	0.0522	3	0.1983	3	0.0559	3	0.4315	3	0.0084	3
	A_5	0.0503	5	0.1913	5	0.0586	1	0.3357	5	0.0054	5
Linear max normalization (Çelen 2014)	A_1	0.2213	4	0.8382	3	0.7886	4	0.4111	4	0.0369	4
	A_2	0.2348	2	0.8741	2	0.837	2	0.5244	2	0.0464	2
	A_3	0.2356	1	0.8815	1	0.8463	1	0.5432	1	0.047	1
	A_4	0.2247	3	0.8371	4	0.7886	3	0.4344	3	0.0415	3
	A_5	0.2164	5	0.8306	5	0.7773	5	0.3213	5	0.026	5
	A_1	0.1006	4	0.3442	4	0.2564	4	0.3442	4	0.1006	4

(Continued)

TABLE 5.17 (Continued)

Ideal solutions		$\{0, \ldots, 0\}$ and $\{1, \ldots, 1\}$ (El Alaoui, El Yassini, & Ben-azza 2019)		$\{0, \ldots, 0\}$ and $\{w_1, \ldots, w_n\}$, (Cables, Lamata, & Verdegay 2016)		$\{w_1 Z_0/Z_1, \ldots, w_i Z_0/Z_i\}$ and $\{w_1, \ldots, w_n\}$, (García-Cascales & Lamata 2012)		$\{\min_i(v_{ij}), \ldots, \min_i(v_{ij})\}$ and $\{\max_i(v_{ij}), \ldots, \max_i(v_{ij})\}$ (Hwang & Yoon 1981)		$\{\min_i(v_{ij}), \ldots, \min_i(v_{ij})\}$ and $\{1, \ldots, 1\}$ (Mukherjee 2014)	
normalization	alternatives	CC_i	rank	CC_i	rank	CC_i	rank	CC_i	rank	CC_i	rank
Linear max-min normalization (Jahan & Edwards 2015)	A_2	0.1787	2	0.5995	2	0.5168	2	0.5995	2	0.1787	2
	A_3	0.1797	1	0.6078	1	0.5241	1	0.6078	1	0.1797	1
	A_4	0.1366	3	0.4263	3	0.3611	3	0.4263	3	0.1366	3
	A_5	0.0764	5	0.2673	5	0.1952	5	0.2673	5	0.0764	5
Logarithmic normalization (Zavadskas, Ustinovichius, & Peldschus 2003)	A_1	0.3307	4	0.818	2	0.7852	2	0.4444	3	0.0186	4
	A_2	0.3349	2	0.8072	5	0.7737	5	0.5189	2	0.0212	2
	A_3	0.3354	1	0.8077	4	0.7743	4	0.5514	1	0.0216	1
	A_4	0.3313	3	0.8118	3	0.7784	3	0.4271	4	0.019	3
	A_5	0.3293	5	0.8187	1	0.7858	1	0.3586	5	0.0135	5
Markovic normalization (Marković 2010)	A_1	0.2398	2	0.8902	2	0.8577	2	0.5889	2	0.0528	3
	A_2	0.2263	4	0.8569	4	0.8127	4	0.4756	4	0.0424	4
	A_3	0.2253	5	0.8552	5	0.8103	5	0.4568	5	0.0398	5
	A_4	0.2372	3	0.8763	3	0.8399	3	0.5656	3	0.0541	2
	A_5	0.2437	1	0.9217	1	0.8978	1	0.6787	1	0.0547	1
Tzeng and Huang normalization	A_1	0.3062	2	0.8507	4	0.8199	4	0.5485	3	0.0719	2
	A_2	0.2924	4	0.8652	2	0.8348	2	0.4786	4	0.0605	4
	A_3	0.2898	5	0.8773	1	0.8488	1	0.4432	5	0.054	5

(Continued)

TABLE 5.17 (Continued)

Ideal solutions	{0,...,0} and {1,...,1} (El Alaoui, El Yassini, & Benazza 2019)		{0,...,0} and {w_1,...,w_n} (Cables, Lamata, & Verdegay 2016)		{$w_1 Z_0/Z_1$,...,$w_n Z_0/Z_1$} and {w_1,...,w_n} (García-Cascales & Lamata 2012)		{$\min_i(v_{ij})$,...,$\min_i(v_{ij})$} and {$\max_i(v_{ij})$,...,$\max_i(v_{ij})$} (Hwang & Yoon 1981)		{$\min_i(v_{ij})$,...,$\min_i(v_{ij})$} and {1,...,1} (Mukherjee 2014)	
normalization alternatives	CC_i	rank	CC_i	rank	CC_i	rank	CC_i	rank	CC_i	rank
(Tzeng & Huang 2011)										
A_4	0.3055	3	0.839	5	0.8064	5	0.5586	2	0.0778	1
A_5	0.3095	1	0.8534	3	0.8234	3	0.6038	1	0.0715	3
Nonlinear normalization (Peldschus, Vaigauskas, & Zavadskas 1983)										
A_1	0.1913	4	0.713	4	0.6282	4	0.3922	4	0.0579	4
A_2	0.2172	2	0.7855	2	0.7266	2	0.5242	2	0.0771	2
A_3	0.2181	1	0.794	1	0.7366	1	0.5376	1	0.0776	1
A_4	0.1994	3	0.7223	3	0.6453	3	0.429	3	0.0671	3
A_5	0.1811	5	0.6917	5	0.5953	5	0.2875	5	0.0383	5
Lai and Hwang normalization (Lai & Hwang 1994)										
A_1	0.6807	2	0.5702	2	0.5552	3	0.3442	4	0.1725	4
A_2	0.6592	5	0.5619	5	0.5484	5	0.5995	2	0.2398	2
A_3	0.661	4	0.562	4	0.5485	4	0.6078	1	0.2415	1
A_4	0.6754	3	0.5676	3	0.553	2	0.4263	3	0.2167	3
A_5	0.6822	1	0.5721	1	0.5567	1	0.2673	5	0.1389	5
Zavadskas and Turskis (Zavadskas & Turskis 2008)										
A_1	0.2213	4	0.8382	3	0.7886	4	0.4111	4	0.0369	4
A_2	0.2348	2	0.8741	2	0.837	2	0.5244	2	0.0464	2
A_3	0.2356	1	0.8815	1	0.8463	1	0.5432	1	0.047	1
A_4	0.2247	3	0.8371	4	0.7886	3	0.4344	3	0.0415	3
A_5	0.2164	5	0.8306	5	0.7773	5	0.3213	5	0.026	5

Linear max normalization—which García-Cascales and Lamata (2012) use with $\{w_1 Z_0/Z_t, \ldots, w_n Z_0/Z_t\}$ and $\{w_1, \ldots, w_n\}$ to avoid rank reversal—and Zavadskas and Turskis normalization result in rank permutation with the ideal solutions $\{0, 0, \ldots, 0\}$ and $\{w_1, \ldots, w_n\}$ between A_1 and A_4, ranked second and third.

Even if Marković normalization was proposed for benefit criteria, which is the case in this example, it results in total rank reversal with all sets of ideal solutions, which suggests modifying the formula to $r_{ij} = (x_{ij} - x_j^-)/x_j^*$.

The remaining normalizations all result in total rank reversal with some ideal solutions and the correct rank with others.

DISTANCE OR METRIC USED

Since the initial introduction of TOPSIS (Hwang & Yoon, 1981), the influence on the final result of the distance used has been apparent. For example, using the city-block distance (called also Manhattan distance) inverts the ranks of A_1 and A_3 in Table 5.1.

Olson (2004) compares the Manhattan distance L_1, the Euclidean distance L_2, and the Chebyshev distance L_∞ and concludes that the Chebyshev metric is slightly less adequate than the other two for TOPSIS. He also mentions that the performance of Manhattan and Euclidean distances are comparable.

In addition to these metrics, Shih, Shyur, and Lee (2007) use the weighted L_p metric. Similarly, they conclude that Manhattan and Euclidean distances are more suitable.

Without taking a position in favor of one distance metric over another, Ploskas and Papathanasiou (2019) present 15 metrics in conjunction with TOPSIS. Using the example in Table 5.8 and keeping the different parameters of the original TOPSIS method, Table 5.18 summarizes the definitions of distances used by Ploskas and Papathanasiou (2019), the values obtained, and the rankings generated.

All distance metrics except the Pearson distance rank A_3 first. This is due to the dominator term containing exclusively reference points. All 15 distance metrics rank A_5 last.

Even if the Manhattan and Euclidean distances may produce different results (Hwang & Yoon, 1981), in this example they lead to the same ranking, along with the squared Euclidean, Sørensen, Lorentzian, Jaccard, and Dice distances. The rank produced by these seven distances is called rank 1 (Table 5.19), and that obtained using the Chebyshev, Bhattacharyya, Hellinger, and Matusita distances—with one permutation, between A_1 and A_4 in third and fourth place—is called rank 2.

There are also three other less-consensual rankings: the Canberra distance is one permutation away from rank 1: between A_1 and A_2 in second and fourth. The ranks produced by the squared-chord and squared Pearson distances are also one permutation away from rank 2—between A_1 and A_2 in second and third—along with, in the case of the Pearson distance, one permutation away from the rank of the Canberra distance (between A_1 and A_3 in first and second).

As pointed out by Ploskas and Papathanasiou (2019), the choice of distance must be adapted to the situation being treated. The combination of the absolute reference point $\{0, 0, \ldots, 0\}$ and the Pearson distance is mathematically impossible, and that

TABLE 5.18
Effect of Distance Metrics on Rank Reversal

Distance Metric	Alternative	D_i^* Definition	Value	D_i^- Definition	Value	CC_i	Rank
Manhattan (Jiang et al., 2019)	A_1	$\sum_{j=1}^{n} \lvert v_{ij} - v_j^* \rvert$	0.0772	$\sum_{j=1}^{n} \lvert v_{ij} - v_j^- \rvert$	0.0476	0.3814	4
	A_2		0.0629		0.0619	0.4959	2
	A_3		0.0601		0.0647	0.5188	1
	A_4		0.0746		0.0502	0.4021	3
	A_5		0.0857		0.0391	0.3135	5
Euclidean (Arslan, 2017)	A_1	$\sqrt{\sum_{j=1}^{n} (v_{ij} - v_j^*)^2}$	0.0447	$\sqrt{\sum_{j=1}^{n} (v_{ij} - v_j^-)^2}$	0.0332	0.4261	4
	A_2		0.0374		0.0388	0.5086	2
	A_3		0.0349		0.0388	0.5262	1
	A_4		0.0469		0.0356	0.4317	3
	A_5		0.0455		0.0229	0.3348	5
Chebyshev	A_1	$max(\lvert v_{ij} - v_j^* \rvert)$	0.0271	$min(\lvert v_{ij} - v_j^- \rvert)$	0.0309	0.5334	3
	A_2		0.0284		0.0338	0.5437	2
	A_3		0.0225		0.033	0.5943	1
	A_4		0.0338		0.0271	0.4448	4
	A_5		0.033		0.0155	0.3189	5
Squared Euclidean	A_1	$\sum_{j=1}^{n} (v_{ij} - v_j^*)^2$	0.002	$\sum_{j=1}^{n} (v_{ij} - v_j^-)^2$	0.0011	0.3553	4
	A_2		0.0014		0.0015	0.5172	2
	A_3		0.0012		0.0015	0.5522	1
	A_4		0.0022		0.0013	0.3658	3
	A_5		0.0021		0.0005	0.2021	5
Sørensen (Sørensen, 1948) or Bray–Curtis (Bray & Curtis, 1957)	A_1		0.0807		0.0572	0.4149	4

(Continued)

TABLE 5.18 (Continued)

Distance Metric	Alternative	D_i^* Definition	Value	D_i^- Definition	Value	CC_i	Rank
	A_2	$\frac{\sum_{j=1}^{n}|v_{ij}-v_j^*|}{\sum_{j=1}^{n}(v_{ij}+v_j^*)}$	0.0648	$\frac{\sum_{j=1}^{n}|v_{ij}-v_j^-|}{\sum_{j=1}^{n}(v_{ij}+v_j^-)}$	0.0731	0.5303	2
	A_3		0.0617		0.0762	0.5529	1
	A_4		0.0778		0.0601	0.436	3
	A_5		0.0903		0.0475	0.3447	5
Canberra (Lance & Williams, 1966)	A_1	$\sum_{j=1}^{n}\frac{|v_{ij}-v_j^*|}{(v_{ij}+v_j^*)}$	0.3387	$\sum_{j=1}^{n}\frac{|v_{ij}-v_j^-|}{(v_{ij}+v_j^-)}$	0.2605	0.4348	2
	A_2		0.3441		0.257	0.4276	4
	A_3		0.3205		0.2813	0.4675	1
	A_4		0.342		0.2569	0.429	3
	A_5		0.3584		0.244	0.4051	5
Lorentzian (Deza & Deza, 2006)	A_1	$\sum_{j=1}^{n}\ln(1+|v_{ij}-v_j^*|)$	0.0762	$\sum_{j=1}^{n}\ln(1+|v_{ij}-v_j^-|)$	0.0471	0.3817	4
	A_2		0.0622		0.0612	0.4957	2
	A_3		0.0595		0.064	0.5184	1
	A_4		0.0735		0.0496	0.4026	3
	A_5		0.0847		0.0389	0.3147	5
Jaccard (Jaccard, 1902)	A_1	$\frac{\sum_{j=1}^{n}(v_{ij}-v_j^*)^2}{\sum_{j=1}^{n}(v_{ij}+v_j^*)^2-v_{ij}v_j^*}$	0.0313	$\frac{\sum_{j=1}^{n}(v_{ij}-v_j^-)^2}{\sum_{j=1}^{n}(v_{ij}+v_j^-)^2-v_{ij}v_j^-}$	0.0226	0.4187	4
	A_2		0.0209		0.0286	0.5774	2
	A_3		0.0182		0.0286	0.6109	1
	A_4		0.034		0.0256	0.429	3
	A_5		0.0332		0.0111	0.2511	5
Dice (Dice, 1945)	A_1	$\frac{\sum_{j=1}^{n}(v_{ij}-v_j^*)^2}{\sum_{j=1}^{n}v_{ij}^2+\sum_{j=1}^{n}v_j^{*2}}$	0.0159	$\frac{\sum_{j=1}^{n}(v_{ij}-v_j^-)^2}{\sum_{j=1}^{n}v_{ij}^2+\sum_{j=1}^{n}v_j^{-2}}$	0.0114	0.4176	4
	A_2		0.0106		0.0145	0.5783	2
	A_3		0.0092		0.0145	0.6121	1

(Continued)

TABLE 5.18 (Continued)

Distance Metric	Alternative	D_i^* Definition	Value	D_i^- Definition	Value	CC_i	Rank
	A_4		0.0173		0.0129	0.4279	3
	A_5		0.0169		0.0056	0.249	5
Bhattacharyya (Bhattacharyya, 1946)	A_1	$-ln\sum_{j=1}^{n}\sqrt{v_{ij}v_j^*}$	1.4838	$-ln\sum_{j=1}^{n}\sqrt{v_{ij}v_j^-}$	1.7621	0.5429	3
	A_2		1.4531		1.7263	0.543	2
	A_3		1.4459		1.7198	0.5433	1
	A_4		1.4799		1.7558	0.5426	4
	A_5		1.5007		1.7786	0.5424	5
Hellinger (Deza & Deza, 2006)	A_1	$2\sqrt{1-\sum_{j=1}^{n}\sqrt{v_{ij}v_j^*}}$	1.4475	$2\sqrt{1-\sum_{j=1}^{n}\sqrt{v_{ij}v_j^-}}$	1.5306	0.514	3
	A_2		1.4373		1.5207	0.5141	2
	A_3		1.4348		1.5189	0.5142	1
	A_4		1.4462		1.5288	0.5139	4
	A_5		1.453		1.535	0.5137	5
Matusita (Matusita, 1955)	A_1	$\sqrt{2-2\sum_{j=1}^{n}\sqrt{v_{ij}v_j^*}}$	1.0235	$\sqrt{2-2\sum_{j=1}^{n}\sqrt{v_{ij}v_j^-}}$	1.0823	0.514	3
	A_2		1.0163		1.0753	0.5141	2
	A_3		1.0146		1.0741	0.5142	1
	A_4		1.0226		1.0811	0.5139	4
	A_5		1.0274		1.0854	0.5137	5
Squared-chord (Prentice, 1980)	A_1	$\sum_{j=1}^{n}(\sqrt{v_{ij}}-\sqrt{v_j^*})^2$	0.0045	$\sum_{j=1}^{n}(\sqrt{v_{ij}}-\sqrt{v_j^-})^2$	0.0034	0.4319	2
	A_2		0.004		0.0027	0.4016	3
	A_3		0.0034		0.0028	0.4559	1
	A_4		0.0052		0.0033	0.3928	4
	A_5		0.004		0.0017	0.2991	5

(Continued)

TABLE 5.18 (Continued)

Distance Metric	Alternative	D_i^* Definition	Value	D_i^- Definition	Value	CC_i	Rank
Pearson (Pearson, 1900)	A_1	$\sum_{j=1}^{n} \frac{(v_{ij}-v_j^*)^2}{v_j^*}$	0.0157	$\sum_{j=1}^{n} \frac{(v_{ij}-v_j^-)^2}{v_j^-}$	0.0165	0.5135	1
	A_2		0.0139		0.0119	0.4609	4
	A_3		0.0119		0.0124	0.5103	2
	A_4		0.0177		0.0156	0.4697	3
	A_5		0.0145		0.0075	0.3413	5
Squared (Pearson, 1900)	A_1	$\sum_{j=1}^{n} \frac{(v_{ij}-v_j^*)^2}{v_{ij}+v_j^*}$	0.0089	$\sum_{j=1}^{n} \frac{(v_{ij}-v_j^-)^2}{v_{ij}+v_j^-}$	0.0067	0.4308	2
	A_2		0.008		0.0054	0.4024	3
	A_3		0.0067		0.0056	0.4563	1
	A_4		0.0102		0.0066	0.3931	4
	A_5		0.008		0.0034	0.2991	5

TABLE 5.19
Correlations among the Five Rankings Obtained

Rankings	Rank 1	Rank 2	Canberra	Squared-Chord	Pearson
Rank 1	1	0.9	0.6	0.7	0.3
Rank 2	0.9	1	0.7	0.9	0.5
Canberra	0.6	0.7	1	0.9	0.9
Squared chored	0.7	0.9	0.9	1	0.8
Pearson	0.3	0.5	0.9	0.8	1

reference point is completely senseless when combined with the Matusita, Hellinger, and Bhattacharya distances. However, some rankings produced are less acceptable than others (although this is not due solely to the distance used). For example, it is difficult to accept the Pearson-distance ranking, because it favors A_1, an alternative that is dominated by A_2 and A_3 according to three of the four criteria. For the same reasons, the rankings of the Canberra, squared-chord, and squared Pearson distances are more than questionable. Here one would prefer the ranking produced by the Manhattan, Euclidean, squared Euclidean, Sørensen, Lorentzian, Jaccard, and Dice distances, because it is identical for three combinations of reference points among the five studied, including the one avoiding rank reversal. Furthermore, knowing that C_2 and C_3 have the same weight (1/3) and C_1 and C_4 also have the same weight (1/6), the gap by which A_4 outperforms A_1 according to C_3 is bigger than the gap by which A_1 outperforms A_4 according to C_2. The same could be said in favor of A_4 compared to A_1 with regard to C_1 and C_4.

CONCLUSION

This chapter reviews the original TOPSIS algorithm and proposes a relevant MATLAB algorithm. It maintains that like other methods, TOPSIS suffers rank reversal. Thus, the chapter reviews the main causes treated in the literature: ideal solutions and normalization methods.

Since no proposed approach that deals with just one aspect (ideal solutions or normalization method) can totally prevent rank reversal, the chapter discusses five interpretations of ideal solutions and 10 normalization methods and shows experimentally, without pretending to find the magical solution, that some combinations are more likely to cause rank reversal than others. It also discusses 15 distance metrics and their effects on the final ranking.

REFERENCES

Acuña-Soto, C., V. Liern, & B. Pérez-Gladish (2018). "Normalization in TOPSIS-Based Approaches with Data of Different Nature: Application to the Ranking of Mathematical Videos." *Annals of Operations Research*, June. https://doi.org/10.1007/s10479-018-2945-5.

Aires, Renan Felinto de Farias, & Luciano Ferreira (2019). "A New Approach to Avoid Rank Reversal Cases in the TOPSIS Method." *Computers & Industrial Engineering*, *132*(June), 84–97. https://doi.org/10.1016/j.cie.2019.04.023.

Aires, Renan Felinto de Farias, Luciano Ferreira, Renan Felinto de Farias Aires, & Luciano Ferreira (2018). "The Rank Reversal Problem in Multi-Criteria Decision Making: A Literature Review." *Pesquisa Operacional*, *38*(2), 331–362. https://doi.org/10.1590/0101-7438.2018.038.02.0331.

Arslan, Turan (2017). "A Weighted Euclidean Distance Based TOPSIS Method for Modeling Public Subjective Judgments." *Asia-Pacific Journal of Operational Research*, *34*(03), 1750004. https://doi.org/10.1142/S021759591750004X.

Barzilai, Jonathan, & Boaz Golany (1994). "AHP Rank Reversal, Normalization And Aggregation Rules." *INFOR: Information Systems and Operational Research*, *32*(2), 57–64. https://doi.org/10.1080/03155986.1994.11732238.

Belton, Valerie, & Tony Gear (1983). "On a Short-Coming of Saaty's Method of Analytic Hierarchies." *Omega*, *11*(3), 228–230. https://doi.org/10.1016/0305-0483(83)90047-6.

Belton, Valerie, & Tony Gear (1985). "The Legitimacy of Rank Reversal—A Comment." *Omega*, *13*(3), 143–144. https://doi.org/10.1016/0305-0483(85)90052-0.

Berghman, Erica, Yves De Smet, Jean Rosenfeld, & Dimitri Van Assche (2019). "A Dichotomous Approach to Reduce Rank Reversal Occurrences in PROMETHEE II Rankings." In Kalyanmoy Deb, Erik Goodman, Carlos A. Coello, Kathrin Klamroth, Kaisa Miettinen, Sanaz Mostaghim, & Patrick Reed edited by, *Evolutionary Multi-Criterion Optimization* (pp. 644–654). Lecture Notes in Computer Science. Cham: Springer International Publishing. https://doi.org/10.1007/978-3-030-12598-1_51.

Bhattacharyya, A. (1946). "On a Measure of Divergence between Two Multinomial Populations." *Sankhyā: The Indian Journal of Statistics (1933-1960)*, *7*(4), 401–406.

Brans, Jean-Pierre, & Yves De Smet (2016). "PROMETHEE Methods." In Salvatore Greco, Matthias Ehrgott, & José Rui Figueira edited by, *Multiple Criteria Decision Analysis: State of the Art Surveys* (pp. 187–219). International Series in Operations Research & Management Science. New York, NY: Springer. https://doi.org/10.1007/978-1-4939-3094-4_6.

Bray, J. Roger, & J. T. Curtis. 1957. "An Ordination of the Upland Forest Communities of Southern Wisconsin." *Ecological Monographs*, *27*(4), 325–349. https://doi.org/10.2307/1942268.

Cables, E., M. T. Lamata, & J. L. Verdegay (2016). "RIM-Reference Ideal Method in Multicriteria Decision Making." *Information sciences*, *337–338* (April), 1–10. https://doi.org/10.1016/j.ins.2015.12.011.

Ceballos, Blanca, David A. Pelta, and María T. Lamata (2017). "Rank Reversal and the VIKOR Method: An Empirical Evaluation." *International Journal of Information Technology & Decision Making*, *17*(02), 513–525. https://doi.org/10.1142/S0219622017500237.

Çelen, Aydın (2014). "Comparative Analysis of Normalization Procedures in TOPSIS Method: With an Application to Turkish Deposit Banking Market." *Informatica*, *25*(2), 185–208.

Chakraborty, S., & C. Yeh (2009). "A Simulation Comparison of Normalization Procedures for TOPSIS." In *2009 International Conference on Computers Industrial Engineering*, 1815–1820. https://doi.org/10.1109/ICCIE.2009.5223811.

Chatterjee, Prasenjit, & Shankar Chakraborty (2014). "Investigating the Effect of Normalization Norms in Flexible Manufacturing System Selection Using Multi-Criteria Decision-Making Methods." *Journal of Engineering Science & Technology Review*, 7(3), 141–150.

Chen, Chen-Tung (2000). "Extensions of the TOPSIS for Group Decision-Making under Fuzzy Environment." *Fuzzy Sets and Systems*, 114(1), 1–9. https://doi.org/10.1016/S0165-0114(97)00377-1.

Chen, Pengyu (2019). "Effects of Normalization on the Entropy-Based TOPSIS Method." *Expert Systems with Applications*, 136(December), 33–41. https://doi.org/10.1016/j.eswa.2019.06.035.

Deza, Michel-Marie, & Elena Deza (2006). *Dictionary of Distances*. Netherlands: Elsevier.

Dice, Lee R. (1945). "Measures of the Amount of Ecologic Association Between Species." *Ecology*, 26(3), 297–302. https://doi.org/10.2307/1932409.

El Alaoui, Mohamed, Hussain Ben-azza, & Khalid El Yassini (2019). "Fuzzy TOPSIS with Coherent Measure: Applied to a Closed Loop Agriculture Supply Chain." In Mostafa Ezziyyani edited by, *Advanced Intelligent Systems for Sustainable Development (AI2SD'2018)*, (pp. 106–117). Advances in Intelligent Systems and Computing. Switzerland: Springer International Publishing. https://doi.org/10.1007/978-3-030-11878-5_12.

El Alaoui, Mohamed, Khalid El Yassini, & Hussain Ben-azza (2019). "Type 2 Fuzzy TOPSIS for Agriculture MCDM Problems." *International Journal of Sustainable Agricultural Management and Informatics*, 5(2/3), 112–130. https://doi.org/10.1504/IJSAMI.2019.101672.

Falkner, Charles H., & Vedat Verter (1991). "Separation and Normalization in Multi-Attribute Decision Models for Investment Evaluation." *The Engineering Economist*, 37(1), 77–85. https://doi.org/10.1080/00137919108903058.

Forman, Ernest H. (1990). "AHP Is Intended for More Than Expected Value Calculations." *Decision Sciences*, 21(3), 670–672. https://doi.org/10.1111/j.1540-5915.1990.tb00343.x.

García-Cascales, M. Socorro, & M. Teresa Lamata (2012). "On Rank Reversal and TOPSIS Method." *Mathematical and Computer Modelling*, 56(5), 123–132. https://doi.org/10.1016/j.mcm.2011.12.022.

Hwang, Ching-Lai, & Kwangsun Yoon (1981). *Multiple Attribute Decision Making: Methods and Applications A State-of-the-Art Survey*. Berlin, Heidelberg: Springer-Verlag. http://www.springer.com/gp/book/9783540105589.

İç, Yusuf Tansel (2014). "A TOPSIS Based Design of Experiment Approach to Assess Company Ranking." *Applied Mathematics and Computation*, 227(January), 630–647. https://doi.org/10.1016/j.amc.2013.11.043.

Jaccard, Paul (1902). "Distribution Comparée de La Flore Alpine Dans Quelques Régions Des Alpes Occidentales et Orientales." *Bulletin de La Murithienne*, 31: 81–92.

Jahan, Ali, & Kevin L. Edwards (2015). "A State-of-the-Art Survey on the Influence of Normalization Techniques in Ranking: Improving the Materials Selection Process in Engineering Design." *Materials & Design (1980-2015)*, 65(January), 335–342. https://doi.org/10.1016/j.matdes.2014.09.022.

Jiang, Wen, Meijuan Wang, Xinyang Deng, & Linfeng Gou (2019). "Fault Diagnosis Based on TOPSIS Method with Manhattan Distance." *Advances in Mechanical Engineering*, 11(3), 1687814019833279. https://doi.org/10.1177/1687814019833279.

Kong, Feng (2011). "Rank Reversal and Rank Preservation in TOPSIS." *Advanced Materials Research*, 204–210. https://doi.org/10.4028/www.scientific.net/AMR.204-210.36.

Kong, Feng, Wei, & Jia-Hao Gong (2016). "Rank Reversal and Rank Preservation in ANP Method." *Journal of Discrete Mathematical Sciences and Cryptography*, 19(3), 821–836. https://doi.org/10.1080/09720529.2016.1197570.

Lai, Young-Jou, & Ching-Lai Hwang (1994). *Fuzzy Multiple Objective Decision Making: Methods and Applications*. Lecture Notes in Economics and Mathematical Systems 404. Berlin, Heidelberg: Springer-Verlag. https://doi.org/10.1007/978-3-642-57949-3.

Lance, G. N., & W. T. Williams (1966). "Computer Programs for Hierarchical Polythetic Classification ('Similarity Analyses')." *The Computer Journal*, 9(1), 60–64. https://doi.org/10.1093/comjnl/9.1.60.

Maleki, Hamed, & Sajjad Zahir (2013). "A Comprehensive Literature Review of the Rank Reversal Phenomenon in the Analytic Hierarchy Process." *Journal of Multi-Criteria Decision Analysis*, 20(3–4), 141–155. https://doi.org/10.1002/mcda.1479.

Mareschal, B., Y. De Smet, & P. Nemery (2008). "Rank Reversal in the PROMETHEE II Method: Some New Results." In *2008 IEEE International Conference on Industrial Engineering and Engineering Management*, 959–963. https://doi.org/10.1109/IEEM.2008.4738012.

Marković, Z. (2010). "Modification of Topsis Method for Solving of Multicriteria Tasks." *Yugoslav Journal of Operations Research*, 20(1), 117–143.

Matusita, Kameo (1955). "Decision Rules, Based on the Distance, for Problems of Fit, Two Samples, and Estimation." *The Annals of Mathematical Statistics*, 26(4), 631–640.

Migilinskas, Darius, & Leonas Ustinovichius (2007). "Normalisation in the Selection of Construction Alternatives." *International Journal of Management and Decision Making*, 8(5), 623.

Milani, A. S., A. Shanian, R. Madoliat, & J. A. Nemes (2005). "The Effect of Normalization Norms in Multiple Attribute Decision Making Models: A Case Study in Gear Material Selection." *Structural and Multidisciplinary Optimization*, 29(4), 312–318.

Millet, Ido, & Thomas L. Saaty (2000). "On the Relativity of Relative Measures – Accommodating Both Rank Preservation and Rank Reversals in the AHP." *European Journal of Operational Research 121*(1), 205–212. https://doi.org/10.1016/S0377-2217(99)00040-5.

Mousavi-Nasab, Seyed Hadi, & Alireza Sotoudeh-Anvari (2018). "A New Multi-Criteria Decision Making Approach for Sustainable Material Selection Problem: A Critical Study on Rank Reversal Problem." *Journal of Cleaner Production*, 182(May), 466–484. https://doi.org/10.1016/j.jclepro.2018.02.062.

Mufazzal, Sameera, & S. M. Muzakkir (2018). "A New Multi-Criterion Decision Making (MCDM) Method Based on Proximity Indexed Value for Minimizing Rank Reversals." *Computers & Industrial Engineering*, 119(May), 427–438. https://doi.org/10.1016/j.cie.2018.03.045.

Mukherjee, Krishnendu (2014). "Analytic Hierarchy Process and Technique for Order Preference by Similarity to Ideal Solution: A Bibliometric Analysis 'from' Past, Present and Future of AHP and TOPSIS." *International Journal of Intelligent Engineering Informatics*, 2(2–3), 96–117. https://doi.org/10.1504/IJIEI.2014.066210.

Olson, D. L. (2004). "Comparison of Weights in TOPSIS Models." *Mathematical and Computer Modelling*, 40(7), 721–727. https://doi.org/10.1016/j.mcm.2004.10.003.

Pavličić, Dubravka M. (2001). "Normalization Affects the Results of MADM Methods." *Yugoslav Journal of Operations Research*, 11(2), 251–265.

Pearson, Karl (1900). "On the Criterion That a given System of Deviations from the Probable in the Case of a Correlated System of Variables Is Such That It Can Be Reasonably Supposed to Have Arisen from Random Sampling." *The London, Edinburgh, and Dublin Philosophical Magazine and Journal of Science*, 50(302), 157–175. https://doi.org/10.1080/14786440009463897.

Peldschus, F., E. Vaigauskas, & E. K. Zavadskas (1983). "Technologische Entscheidungen Bei Der Berücksichtigung Mehrerer Ziehle." *Bauplanung Bautechnik*, 37(4), 173–175.

Pérez, Joaquin (1995). "Some Comments on Saaty's AHP." *Management Science*, 41(6), 1091–1095. https://doi.org/10.1287/mnsc.41.6.1091.

Ploskas, Nikolaos, & Jason Papathanasiou (2019). "A Decision Support System for Multiple Criteria Alternative Ranking Using TOPSIS and VIKOR in Fuzzy and Nonfuzzy Environments." *Fuzzy Sets and Systems*, Theme: Preference, Decision, Optimization, *377*(December), 1–30. https://doi.org/10.1016/j.fss.2019.01.012.

Prentice, I. C. (1980). "Multidimensional Scaling as a Research Tool in Quaternary Palynology: A Review of Theory and Methods." *Review of Palaeobotany and Palynology*, *31*(January), 71–104. https://doi.org/10.1016/0034-6667(80)90023-8.

Ramanathan, Usha, & Ramakrishnan Ramanathan (2011). "An Investigation into Rank Reversal Properties of the Multiplicative AHP." *International Journal of Operational Research*, *11*(1), 54–77. https://doi.org/10.1504/IJOR.2011.040328.

Saaty, R. W. (1987). "The Analytic Hierarchy Process—What It Is and How It Is Used." *Mathematical Modelling*, *9*(3), 161–176. https://doi.org/10.1016/0270-0255(87)90473-8.

Saaty, Thomas L. (1990). "An Exposition on the AHP in Reply to the Paper 'Remarks on the Analytic Hierarchy Process.'" *Management Science*, *36*(3), 259–268.

Saaty, Thomas L., & Mujgan Sagir (2009). "An Essay on Rank Preservation and Reversal." *Mathematical and Computer Modelling*, *49*(5), 1230–1243. https://doi.org/10.1016/j.mcm.2008.08.001.

Saaty, Thomas L, & Luis G. Vargas (1984). "The Legitimacy of Rank Reversal." *Omega*, *12*(5), 513–516. https://doi.org/10.1016/0305-0483(84)90052-5.

Sarraf, Amin Zadeh, Ali Mohaghar, & Hossein Bazargani (2013). "Developing TOPSIS Method Using Statistical Normalization for Selecting Knowledge Management Strategies." *Journal of Industrial Engineering and Management*, *6*(4), 860–875. https://doi.org/10.3926/jiem.573.

Schenkerman, Stan (1994). "Avoiding Rank Reversal in AHP Decision-Support Models." *European Journal of Operational Research*, *74*(3), 407–419. https://doi.org/10.1016/0377-2217(94)90220-8.

Senouci, Mohamed Abdelkrim, M. Sajid Mushtaq, Said Hoceini, & Abdelhamid Mellouk (2016). "TOPSIS-Based Dynamic Approach for Mobile Network Interface Selection." *Computer Networks*, Mobile Wireless Networks, *107*(October), 304–314. https://doi.org/10.1016/j.comnet.2016.04.012.

Shih, Hsu-Shih, Huan-Jyh Shyur, & E. Stanley Lee (2007). "An Extension of TOPSIS for Group Decision Making." *Mathematical and Computer Modelling*, *45*(7), 801–813. https://doi.org/10.1016/j.mcm.2006.03.023.

Sørensen, Thorvald J. (1948). *A Method of Establishing Groups of Equal Amplitude in Plant Sociology Based on Similarity of Species Content and Its Application to Analyses of the Vegetation on Danish Commons*. Munksgaard: I kommission hos E.

Tang, Houxing, & Fang (2018). "A Novel Improvement on Rank Reversal in TOPSIS Based on the Efficacy Coefficient Method." *International Journal of Internet Manufacturing and Services*, *5*(1), 67–84. https://doi.org/10.1504/IJIMS.2018.090591.

Tzeng, Gwo-Hshiung, & Jih-Jeng Huang (2011). *Multiple Attribute Decision Making: Methods and Applications*. USA: CRC Press.

Vafaei, Nazanin, Rita A. Ribeiro, & Luis M. Camarinha-Matos (2018). "Data Normalisation Techniques in Decision Making: Case Study with TOPSIS Method." *International Journal of Information and Decision Sciences*, *10*(1), 19–38. https://doi.org/10.1504/IJIDS.2018.090667.

Vargas, Luis G. (1994). "Reply to Schenkerman's Avoiding Rank Reversal in AHP Decision Support Models." *European Journal of Operational Research*, *74*(3), 420–425. https://doi.org/10.1016/0377-2217(94)90221-4.

Verly, Céline, & Yves De Smet (2013). "Some Results about Rank Reversal Instances in the PROMETHEE Methods." *International Journal of Multicriteria Decision Making 71* 3 (4): 325–345.

Wang, Xiaoting, & Evangelos Triantaphyllou. 2008. "Ranking Irregularities When Evaluating Alternatives by Using Some ELECTRE Methods." *Omega*, Special Issue Section: Papers presented at the INFORMS conference, Atlanta, 2003, *36*(1), 45–63. https://doi.org/10.1016/j.omega.2005.12.003.

Wang, Ying-Ming, & Taha M. S. Elhag (2006). "An Approach to Avoiding Rank Reversal in AHP." *Decision Support Systems*, *42*(3), 1474–1480. https://doi.org/10.1016/j.dss.2005.12.002.

Wang, Ying-Ming, & Ying Luo (2009). "On Rank Reversal in Decision Analysis." *Mathematical and Computer Modelling*, *49*(5), 1221–1229. https://doi.org/10.1016/j.mcm.2008.06.019.

Wijnmalen, Diederik J. D., & William C. Wedley (2008). "Correcting Illegitimate Rank Reversals: Proper Adjustment of Criteria Weights Prevent Alleged AHP Intransitivity." *Journal of Multi-Criteria Decision Analysis*, *15*(5–6), 135–141. https://doi.org/10.1002/mcda.431.

Yang, Wenguang (2020). "Ingenious Solution for the Rank Reversal Problem of TOPSIS Method." *Mathematical Problems in Engineering*, (no. 2): 1–12. https://doi.org/10.1155/2020/9676518.

Yu, L., Y. Pan, & Y. Wu (2009). "Research on Data Normalization Methods in Multi-Attribute Evaluation." In *2009 International Conference on Computational Intelligence and Software Engineering*, 1–5. https://doi.org/10.1109/CISE.2009.5362721.

Zaidan, B. B., & A. A. Zaidan (2018). "Comparative Study on the Evaluation and Benchmarking Information Hiding Approaches Based Multi-Measurement Analysis Using TOPSIS Method with Different Normalisation, Separation and Context Techniques." *Measurement*, *117*(March), 277–294. https://doi.org/10.1016/j.measurement.2017.12.019.

Zanakis, Stelios H., Anthony Solomon, Nicole Wishart, & Sandipa Dublish (1998). "Multi-Attribute Decision Making: A Simulation Comparison of Select Methods." *European Journal of Operational Research*, *107*(3), 507–529. https://doi.org/10.1016/S0377-2217(97)00147-1.

Zavadskas, Edmundas Kazimieras, & Zenonas Turskis (2008). "A New Logarithmic Normalization Method in Games Theory." *Informatica*, *19*(2), 303–314.

Zavadskas, Edmundas Kazimieras, Leonas Ustinovichius, & Friedel Peldschus (2003). "Development of Software for Multiple Criteria Evaluation." *Informatica*, *14*(2), 259–272.

Zavadskas, Edmundas Kazimieras, Algimantas Zakarevicius, & Jurgita Antucheviciene. 2006. "Evaluation of Ranking Accuracy in Multi-Criteria Decisions." *Informatica*, *17*(4), 601–618.

6 Fuzzy TOPSIS

NEGI'S APPROACH

The Approach

The criteria used in the decision-making process can be categorized as quantitative or qualitative. While the former depend only on the measured reality, the latter depend heavily on the assessment by the person in charge. Thus, in order to reduce or even eliminate this bias, assessing alternatives according to qualitative criteria often involves several decision makers.

Humans' assessments are more likely and naturally expressed by linguistic variables such as good, fair, or poor than by crisp numbers. This is why fuzzy logic is doubly useful in this kind of situation. On the one hand, it allows a better representation of linguistic variables, closing the gap between human and machine languages (El Alaoui & El Yassini, 2020). On the other hand, crisp numbers—classically used for quantitative criteria—can be seen as a special case of fuzzy numbers (see Chapter 3). Fuzzy logic therefore seems to be the perfect tool to model situations including both quantitative and qualitative criteria. A search in the Scopus database performed on April 10, 2020, for the terms ("fuzzy" AND "TOPSIS") OR ("fuzziness" AND "TOPSIS") produced 3,213 results for the period 1987–2019, as shown in Figure 6.1.

Chen and Hwang (1992) discuss extensions of several multiattribute decision-making method into the fuzzy context, including simple additive weighting, the analytic hierarchy process, ELECTRE, and PROMETHEE, with perhaps the first fuzzy TOPSIS algorithm based on Negi's work (Negi, 1989). As in crisp TOPSIS (Chapter 5), the algorithm required alternative assessments according to each criterion $X = (x_{ij})_{m \times n}$ and criterion weight $W = (w_j)_{1 \times n}$. However, the assessments x_{ij} and w_j could be either crisp or fuzzy values. For fuzzy values, Negi opts for trapezoidal fuzzy numbers, such that $\tilde{x}_{ij} = (x_{ij}^1, x_{ij}^2, x_{ij}^3, x_{ij}^4)$ and $\tilde{w}_j = (w_j^1, w_j^2, w_j^3, w_j^4)$. Since crisp values are special cases of fuzzy values, in the remainder of this section all values will be considered trapezoidal fuzzy numbers unless stated otherwise.

Thus, the algorithm is as follows:

Algorithm 6.1:

Step 1: Compute the normalized decision matrix $\tilde{R} = [\tilde{r}_{ij}]_{m \times n}$.

For benefit criteria,

$$\tilde{r}_{ij} = \left(\frac{x_{ij}^1}{x_j^{4*}}, \frac{x_{ij}^2}{x_j^{3*}}, \frac{x_{ij}^3}{x_j^{2*}}, \frac{x_{ij}^4}{x_j^{1*}} \right), \qquad (6.1)$$

with $x_j^{1*} = \max_i x_{ij}^1$, $x_j^{2*} = \max_i x_{ij}^2$, $x_j^{3*} = \max_i x_{ij}^3$ and $x_j^{4*} = \max_i x_{ij}^4$;

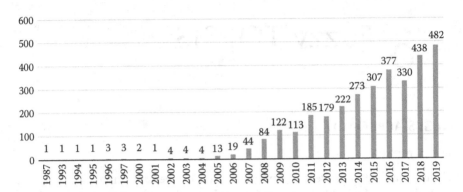

FIGURE 6.1 Fuzzy TOPSIS in the Scopus Database, 1987–2019.

For cost criteria,

$$\tilde{r}_{ij} = \left(\frac{x_j^{1-}}{x_{ij}^4}, \frac{x_j^{2-}}{x_{ij}^3}, \frac{x_j^{3-}}{x_{ij}^2}, \frac{x_j^{4-}}{x_{ij}^1} \right), \quad (6.2)$$

with $x_j^{1-} = \min_i x_{ij}^1$, $x_j^{2-} = \min_i x_{ij}^2$, $x_j^{3-} = \min_i x_{ij}^3$ and $x_j^{4-} = \min_i x_{ij}^4$.

Step 2: Compute the weighted normalized decision matrix $\tilde{V} = [\tilde{v}_{ij}]_{m \times n}$.

$$\tilde{v}_{ij} = \tilde{r}_{ij} \otimes \tilde{w}_j \quad (6.3)$$

For positive trapezoidal fuzzy values, fuzzy multiplication can be computed by Equation 3.13:

$$v_{ij}^1 = r_{ij}^1 * w_j^1, \ v_{ij}^2 = r_{ij}^2 * w_j^2, \ v_{ij}^3 = r_{ij}^3 * w_j^3 \text{ and } v_{ij}^4 = r_{ij}^4 * w_j^4.$$

Step 3: Compute the ideal solutions.

For noncrisp values, Negi opts for an early defuzzification based on the ranking function proposed by Lee and Li (1988), as follows:

$$v_{ij} = \mathcal{R}(\tilde{v}_{ij}) = \frac{-v_{ij}^{1^2} - v_{ij}^{2^2} + v_{ij}^{3^2} + v_{ij}^{4^2} - v_{ij}^1 * v_{ij}^2 + v_{ij}^3 * v_{ij}^4}{3(-v_{ij}^1 - v_{ij}^2 + v_{ij}^3 + v_{ij}^4)}. \quad (6.4)$$

Then the ideal solutions are calculated as in crisp TOPSIS.

The positive ideal solution is

Fuzzy TOPSIS

$$A^* = \{v_1^*, \ldots, v_n^*\} = \max_i v_{ij}, \quad (6.5)$$

and the negative ideal solution is

$$A^- = \{v_1^-, \ldots, v_n^-\} = \min_i v_{ij}. \quad (6.6)$$

Due to the normalization used (Equations 6.1 and 6.2), all weighted normalized assessments are to be considered as benefit criteria.

Step 4: Compute the distances to the ideal solutions.

For crisp values,

$$D_i^* = \sum_j^n |v_{ij} - v_j^*| \quad (6.7)$$

$$D_i^- = \sum_j^n |v_{ij} - v_j^-|. \quad (6.8)$$

For fuzzy values,

$$D_i^* = \sum_j^n \sup_x \left(\mu_{v_{ij}}(x), \mu_{v_j^*}(x) \right) \quad (6.9)$$

$$D_i^- = \sum_j^n \sup_x \left(\mu_{v_{ij}}(x), \mu_{v_j^-}(x) \right). \quad (6.10)$$

Step 5: Calculate the closeness coefficients and rank accordingly.

The formula for the closeness coefficient is

$$CC_i = D_i^- / (D_i^- + D_i^*). \quad (6.11)$$

EXAMPLE

The problem of selecting a fighter jet presented by Chen and Hwang (1992) has four alternatives—models A_1, A_2, A_3, and A_4—evaluated according to six criteria—C_1: maximum speed; C_2: ferry range; C_3: maximum payload; C_4: acquisition cost; C_5: reliability; and C_6: maneuverability. C_1, C_2, C_3, and C_4 are quantitative criteria expressed by crisp numbers, but C_5 and C_6 are qualitative criteria expressed by fuzzy numbers. All criteria except C_4 are benefit criteria. Table 6.1 details the evaluation matrix.

Step 1: Compute the normalized decision matrix.

As in El Alaoui (2018), the tuples of the chosen fuzzy numbers are in the unity interval [0, 1], and thus criteria C_5 and C_6 do not require any normalization. Hence, using Equation 6.1 for benefit criteria with crisp values (C_1, C_2, and C_3) and Equation 6.2 for cost criterion C_4, the normalized decision matrix is given in Table 6.2.

TABLE 6.1
Decision Matrix for Negi's Approach

Alternative	Criterion					
	C_1	C_2	C_3	C_4	C_5	C_6
A_1	2	1,500	20,000	5.5	(0.3, 0.5, 0.5, 0.7)	(0.9, 0.95, 0.95, 1)
A_2	2.5	2,700	18,000	6.5	(0.1, 0.2, 0.2, 0.3)	(0.3, 0.5, 0.5, 0.7)
A_3	1.8	2,000	21,000	4.5	(0.7, 0.8, 0.8, 0.9)	(0.7, 0.8, 0.8, 0.9)
A_4	2.2	1,800	20,000	5	(0.3, 0.5, 0.5, 0.7)	(0.3, 0.5, 0.5, 0.7)
Criterion weights	(0.6, 0.675, 0.675, 0.75)	(0.4, 0.5, 0.5, 0.6)	(0.4, 0.5, 0.5, 0.6)	(0.4, 0.5, 0.5, 0.6)	(0.75, 0.825, 0.825, 0.9)	(0.9, 0.95, 0.95, 1)

Step 2: Compute the weighted normalized decision matrix.

Using Equations 6.3 and 3.13, the weighted normalized decision matrix is shown in Table 6.3.

Step 3: Compute the ideal solutions.

Using the ranking formula (Equation 6.4), the equivalent crisp weighted normalized decision matrix is given in Table 6.4.

Hence, the positive ideal solution (Equation 6.5) is $A^* = \{0.675, 0.5, 0.5, 0.633, 0.904\}$, and the negative ideal solution (Equation 6.6) is $A^- = \{0.486, 0.275, 0.43, 0.345, 0.17, 0.481\}$.

Step 4: Compute the distances to the ideal solutions.

Using Equations 6.9 and 6.10, the distances to the ideal solutions are given in Table 6.5.

Step 5: Calculate the closeness coefficients and rank accordingly.

Closeness coefficients using Equation 6.11 are also shown in Table 6.5, as is the final ranking.

It is true that the proposed normalization eases the treatment in later steps. However, other ranking methods may produce different rankings. In addition, Chen and Hwang (1992) mention that the method is cumbersome.

CHEN'S APPROACH

THE APPROACH

Further early extensions of TOPSIS into the fuzzy context were proposed by Liang (1999) and Triantaphyllou and Lin (1996).

Triantaphyllou and Lin (1996) tried to adapt fuzzy arithmetic operations to the original TOPSIS method. However, their decision matrix was not normalized in the unity interval [0, 1]. It should be mentioned that this is one of the rare

TABLE 6.2
Normalized Decision Matrix for Negi's Approach

Alternative	Criterion					
	C_1	C_2	C_3	C_4	C_5	C_6
A_1	0.8	0.55	0.95	0.82	(0.3, 0.5, 0.5, 0.7)	(0.9, 0.95, 0.95, 1)
A_2	1	1	0.86	0.69	(0.1, 0.2, 0.2, 0.3)	(0.3, 0.5, 0.5, 0.7)
A_3	0.72	0.74	1	1	(0.7, 0.8, 0.8, 0.9)	(0.7, 0.8, 0.8, 0.9)
A_4	0.88	0.67	0.95	0.9	(0.3, 0.5, 0.5, 0.7)	(0.3, 0.5, 0.5, 0.7)

TABLE 6.3
Weighted Normalized Decision Matrix for Negi's Approach

Alternative	Criterion					
	C_1	C_2	C_3	C_4	C_5	C_6
A_1	(0.48, 0.54, 0.54, 0.6)	(0.22, 0.275, 0.275, 0.33)	(0.38, 0.475, 0.475, 0.57)	(0.328, 0.41, 0.41, 0.492)	(0.225, 0.412, 0.412, 0.63)	(0.81, 0.903, 0.903, 1)
A_2	(0.6, 0.675, 0.675, 0.75)	(0.4, 0.5, 0.5, 0.6)	(0.336, 0.43, 0.43, 0.516)	(0.276, 0.345, 0.345, 0.414)	(0.075, 0.165, 0.165, 0.27)	(0.27, 0.475, 0.475, 0.7)
A_3	(0.432, 0.486, 0.486, 0.54)	(0.296, 0.37, 0.37, 0.444)	(0.4, 0.5, 0.5, 0.6)	(0.4, 0.5, 0.5, 0.6)	(0.428, 0.66, 0.66, 0.81)	(0.63, 0.76, 0.76, 0.9)
A_4	(0.528, 0.594, 0.594, 0.66)	(0.268, 0.335, 0.335, 0.402)	(0.38, 0.475, 0.475, 0.57)	(0.36, 0.45, 0.45, 0.54)	(0.225, 0.412, 0.412, 0.63)	(0.27, 0.475, 0.475, 0.7)

methods—alongside those of Sadi-Nezhad and Damghani (2010) and Wang and Elhag (2006)—that obtain fuzzy closeness coefficients, which means that the closeness coefficients have to be compared to get the final ranking (see Chapter 3).

Without mentioning the word "TOPSIS," Liang (1999) proposed "fuzzy [multicriteria decision making] based on ideal and anti-ideal concepts." The proposed algorithm follows the general architecture of the TOPSIS algorithm.

TABLE 6.4
Equivalent Crisp Weighted Normalized Decision Matrix for Negi's Approach

Alternative	Criterion					
	C_1	C_2	C_3	C_4	C_5	C_6
A_1	0.54	0.275	0.475	0.41	0.422	0.904
A_2	0.675	0.5	0.43	0.345	0.17	0.481
A_3	0.486	0.37	0.5	0.5	0.633	0.763
A_4	0.594	0.335	0.475	0.45	0.422	0.482

TABLE 6.5
Distances to Ideal Solutions, Closeness Coefficients, and Final Ranking for Negi's Approach

Alternative	D_i^-	D_i^*	CC_i	Rank
A_1	2.999	3.174	0.486	2
A_2	2.000	3.293	0.378	4
A_3	3.832	1.318	0.744	1
A_4	3.147	3.505	0.473	3

The first widely recognized proper extension of TOPSIS into the fuzzy context was proposed by Chen (2000). To adapt to the fuzzy context, he also proposed linguistic variables for assessing alternatives and criterion importance. Hence, the algorithm becomes the following:

Algorithm 6.2:

Step 1: Each decision maker (DM) assesses each alternative according to each criterion using a linguistic variable $\tilde{x}_{ijk}$ (Table 6.6).

Step 2: Each DM assesses each criterion's importance using a linguistic variable $\tilde{w}_{jk}$ (Table 6.6).

Step 3: Using the individual alternative assessments $\tilde{x}_{ijk}(x_{ijk}^1, x_{ijk}^2, x_{ijk}^3)$ and criterion weights $\tilde{w}_{jk}(w_{jk}^1, w_{jk}^2, w_{jk}^3)$, compute the collective alternative assessments and criterion weights by

$$\tilde{x}_{ij} = (1/K) * (\tilde{x}_{ij1} \oplus \tilde{x}_{ij2} \oplus \ldots \oplus \tilde{x}_{ijK}) \quad (6.12)$$

Fuzzy TOPSIS

TABLE 6.6
Linguistic Variables for Assessing Alternatives and Criterion Importance

Alternative Assessment	Fuzzy Number	Criterion Importance	Fuzzy Number
Very poor (VP)	(0, 0, 1)	Very low (VL)	(0, 0, 0.1)
Poor (P)	(0, 1, 3)	Low (L)	(0, 0.1, 0.3)
Medium poor (MP)	(1, 3, 5)	Medium low (ML)	(0.1, 0.3, 0.5)
Fair (F)	(3, 5, 7)	Medium (M)	(0.3, 0.5, 0.7)
Medium good (MG)	(5, 7, 9)	Medium high (MH)	(0.5, 0.7, 0.9)
Good (G)	(7, 9, 10)	High (H)	(0.7, 0.9, 1)
Very good (VG)	(9, 10, 10)	Very high (VH)	(0.9, 1, 1)

$$\tilde{w}_j = (1/K) * (\tilde{w}_{j1} \oplus \tilde{w}_{j2} \oplus \ldots \oplus \tilde{w}_{jK}). \quad (6.13)$$

Step 4: Construct the normalized decision matrix $\tilde{R} = [\tilde{r}_{ij}]_{m \times n}$.

For benefit criteria,

$$\tilde{r}_{ij} = \left(\frac{x_{ij}^1}{x_{ij}^*}, \frac{x_{ij}^2}{x_{ij}^*}, \frac{x_{ij}^3}{x_{ij}^*} \right), \quad (6.14)$$

with $x_{ij}^* = \max_i x_{ij}^3$;

for cost criteria,

$$\tilde{r}_{ij} = \left(\frac{x_{ij}^-}{x_{ij}^1}, \frac{x_{ij}^-}{x_{ij}^2}, \frac{x_{ij}^-}{x_{ij}^3} \right), \quad (6.15)$$

with $x_{ij}^- = \min_i x_{ij}^1$.

Step 5: Construct the weighted normalized matrix $\tilde{V} = [\tilde{v}_{ij}]_{m \times n}$.

$$\tilde{v}_{ij} = \tilde{r}_{ij} \otimes \tilde{w}_j \quad (6.16)$$

Step 6: Determine the ideal solutions.

The positive ideal solution is

$$A^* = \{v_1^*, \ldots, v_n^*\} \quad (6.17)$$

with $\tilde{v}_j^* = (1, 1, 1)$,

and the negative ideal solution is

$$A^* = \{\tilde{v}_1^-, \ldots, \tilde{v}_n^-\}, \qquad (6.18)$$

with $\tilde{v}_j^- = (0, 0, 0)$.

Step 7: Compute the distance between each alternative and the ideal solutions.

The distance to the positive ideal solution is

$$D_i^* = \sum_{j=1}^n d(\tilde{v}_{ij}, \tilde{v}_j^*), \ i = 1, \ldots, m, \qquad (6.19)$$

with

$$d(\tilde{v}_{ij}, \tilde{v}_j^*) = \sqrt{\left(\frac{1}{3}\right) * ((v_{ij}^1 - 1)^2 + (v_{ij}^2 - 1)^2 + (v_{ij}^3 - 1)^2)}.$$

The distance to the negative ideal solution is

$$D_i^- = \sum_{j=1}^n d(\tilde{v}_{ij}, \tilde{v}_j^-), \ i = 1, \ldots, m, \qquad (6.20)$$

with

$$d(\tilde{v}_{ij}, \tilde{v}_j^-) = \sqrt{\left(\frac{1}{3}\right) * ((v_{ij}^1)^2 + (v_{ij}^2)^2 + (v_{ij}^3)^2)}.$$

Step 8: Calculate the relative closeness to the ideal solutions.

Similar to Equation 5.7, this step uses the formula

$$CC_i = \frac{D_i^-}{D_i^- + D_i^*}, \ i = 1, \ldots, m. \qquad (6.21)$$

Step 9: Rank according to CC_i values, from highest to lowest. ☐

EXAMPLE

In the example proposed by Chen (2000) (Table 6.7), a software company desires to hire a system analysis engineer from among three alternatives: candidates A_1, A_2, and A_3. The evaluation is performed by three decision makers (DM_1, DM_2, and DM_3) according to five criteria—C_1: emotional steadiness; C_2: oral communication skill; C_3: personality; C_4: past experience; and C_5: self-confidence—using linguistic variables.

Step 1: Each DM assesses each alternative according to each criterion using a linguistic variable $\tilde{x}_{ijk}$ (Table 6.7).
Step 2: Each DM assesses each criterion's importance using a linguistic variable $\tilde{w}_{jk}$ (Table 6.8).
Step 3: Using the individual alternative assessments and criterion weights, compute the collective alternative assessments and criterion weights (Table 6.9).
Step 4: Construct the normalized decision matrix $\tilde{R} = [\tilde{r}_{ij}]_{m \times n}$ (Table 6.10).
Step 5: Construct the weighted normalized matrix $\tilde{V} = [\tilde{v}_{ij}]_{m \times n}$ (Table 6.11).
Step 6: Determine the ideal solutions.

$$A^* = [(1, 1, 1), (1, 1, 1), (1, 1, 1), (1, 1, 1), (1, 1, 1)],$$

$$A^- = [(0, 0, 0), (0, 0, 0), (0, 0, 0), (0, 0, 0), (0, 0, 0)].$$

TABLE 6.7
Linguistic Assessment of Three Alternatives According to Five Criteria by Three Decision Makers

Alternative	Decision Maker	Criterion				
		C_1	C_2	C_3	C_4	C_5
A_1	DM_1	MG	G	F	VG	F
	DM_2	G	MG	G	G	F
	DM_3	MG	F	G	VG	F
A_2	DM_1	G	VG	VG	VG	VG
	DM_2	G	VG	VG	VG	MG
	DM_3	MG	VG	G	VG	G
A_3	DM_1	VG	MG	G	G	G
	DM_2	G	G	MG	VG	G
	DM_3	F	VG	VG	MG	MG

TABLE 6.8
Individual Weight Assessments

Decision Maker	Criterion				
	C_1	C_2	C_3	C_4	C_5
DM_1	H	VH	VH	VH	M
DM_2	VH	VH	H	VH	MH
DM_3	MH	VH	H	VH	MH

TABLE 6.9
Fuzzy Group Decision Matrix for Chen's Approach

Alternative	Criterion				
	C_1	C_2	C_3	C_4	C_5
A_1	(5.67, 7.67, 9.33)	(5, 7, 8.67)	(5.67, 7.67, 9)	(8.33, 9.67, 10)	(3, 5, 7)
A_2	(6.33, 8.33, 9.67)	(9, 10, 10)	(8.33, 9.67, 10)	(9, 10, 10)	(7, 8.67, 9.67)
A_3	(6.33, 8, 9)	(7, 8.67, 9.67)	(7, 8.67, 9.67)	(7, 8.67, 9.67)	(6.33, 8.33, 9.67)
Weights	(0.7, 0.87, 0.97)	(0.9, 1, 1)	(0.77, 0.93, 1)	(0.9, 1, 1)	(0.43, 0.63, 0.83)

TABLE 6.10
Normalized Decision Matrix for Chen's Approach

Alternative	Criterion				
	C_1	C_2	C_3	C_4	C_5
A_1	(0.59, 0.79, 0.97)	(0.5, 0.7, 0.87)	(0.57, 0.77, 0.9)	(0.83, 0.97, 1)	(0.31, 0.52, 0.72)
A_2	(0.66, 0.86, 1)	(0.9, 1, 1)	(0.83, 0.97, 1)	(0.9, 1, 1)	(0.72, 0.9, 1)
A_3	(0.66, 0.83, 0.93)	(0.7, 0.87, 0.97)	(0.7, 0.87, 0.97)	(0.7, 0.87, 0.97)	(0.66, 0.86, 1)

Step 7: Compute the distance between each alternative and the ideal solutions (Table 6.12).
Step 8: Calculate the relative closeness to the ideal solutions (Table 6.12).
Step 9: Rank according to CC_i values, from highest to lowest (Table 6.12).

Despite some computational discordances with Chen (2000), the final ranking remains the same. A comparison with the approach proposed by Yuen (2014) is presented by Madi, Garibaldi, and Wagner (2015).

MATLAB Algorithm

As in the previous chapter, a simple MATLAB code is presented here to adapt TOPSIS to the fuzzy context:

```
% step 0: requirements
p=3; % 3 number of tuples for triangular fuzzy number
% linguistic variables for criterion weights
VL = [0 0 0.1]; % Very Low
L = [0 0.1 0.3]; % Low
ML = [0.1 0.3 0.5]; % Medium Low
M = [0.3 0.5 0.7]; % Medium
```

TABLE 6.11
Weighted Normalized Matrix for Chen's Approach

Alternative	Criterion				
	C_1	C_2	C_3	C_4	C_5
A_1	(0.41, 0.69, 0.93)	(0.45, 0.7, 0.87)	(0.43, 0.72, 0.9)	(0.75, 0.97, 1)	(0.13, 0.33, 0.6)
A_2	(0.46, 0.75, 0.97)	(0.81, 1, 1)	(0.64, 0.9, 1)	(0.81, 1, 1)	(0.31, 0.57, 0.83)
A_3	(0.46, 0.72, 0.9)	(0.63, 0.87, 0.97)	(0.54, 0.81, 0.97)	(0.63, 0.87, 0.97)	(0.28, 0.55, 0.83)

TABLE 6.12
Distances to Ideal Solutions, Closeness Coefficients, and Final Ranking

Alternative	D_i^*	D_i^-	CC_i	Rank
A_1	1.9456	3.4295	0.6380	3
A_2	1.2589	4.1057	0.7653	1
A_3	1.6020	3.7706	0.7018	2

```
MH = [0.5 0.7 0.9]; % Medium High
H = [0.7 0.9 1]; % High
VH = [0.9 1 1]; % Very High
% linguistic variables for assessments
VP = [0 0 1]; % Very Poor
P = [0 1 3]; % Poor
MP = [1 3 5]; % Medium Poor
F = [3 5 7]; % Fair
MG = [5 7 9]; % Medium Good
G = [7 9 10]; % Good
VG = [9 10 10]; % Very Good
% step 1: individual assessment of each alternative by each decision maker
% according to each criterion
XI = [MG G F VG F;
G MG G G F;
MG F G VG F;
G VG VG VG VG;
G VG VG VG MG;
MG VG G VG G;
```

```
VG MG G G G;
G G MG VG G;
F VG VG MG MG];
% step 2: individual assessments of criterion weights by each
decision maker
WI = [H VH VH VH M;
VH VH H VH MH;
MH VH H VH MH];
% step 3: calculate the group decision matrix
[m, n1] = size(XI);
[K, n2]=size(WI); % K: number of decision makers
if n1 == n2 % verify concordance between matrix
n = n1;
end
n = n/p; % n: number of criteria
m = m/K; % m: number of alternatives
W = zeros(1,n*p); % group criterion weights
X = zeros(m,n*p); % group assessments
for j = 1:n*p
W(j) = mean(WI(:,j));
for i = 1:m
X(i,j) = mean(XI(p*i-p+1:p*i,j));
end
end
% step 4: construct the normalized decision matrix
N = zeros(m,n*p); % normalized matrix
BC=[1 1 1 1 1]; % 1:benefit criterion 0:cost criterion
for j = 1:n
MaxC = max(max(X(:,p*j-p+1:p*j)));
minC = min(min(X(:,p*j-p+1:p*j)));
if BC(j) == 1 % for benefit criterion
N(:,p*j-p+1:p*j) = X(:,p*j-p+1:p*j) / MaxC;
else % for cost criterion
N(:,p*j-p+1:p*j) = minC / X(:,p*j-p+1:p*j);
end
end
% step 5: construct the weighted normalized matrix
V = zeros(m,n*p); %weighted normalized matrix
for i = 1:m
for j = 1:n*p
V(i,j) = W(1,j) * N(i,j);
end
end
% step 7: compute distances to ideal solutions
D = zeros(m,2); % D(:,1): to PIS D(:,2): to NIS
d1 = 0;
```

```
d2 = 0;
for i = 1:m
 for j = 1:n
  for q = 0:p-1
   d1 = d1 + (1-V(i,p*j-p+1+q))^2;
   d2 = d2 + V(i,p*j-p+1+q)^2;
  end
  D(i,1) = D(i,1) + sqrt((1/p)*d1);
  D(i,2) = D(i,2) + sqrt((1/p)*d2);
  d1 = 0;
  d2 = 0;
 end
end
% step 8: calculate the closeness coefficients
CC = zeros(m,1); % Closeness Coefficients
for i = 1:m
 CC(i) = D(i,2) / (D(i,1) + D(i,2));
end
% step 9: rank according to CC values
[~,idx] = ismember(CC,sort(CC,'descend')); % idx: rank
```

ALGORITHM COMPONENTS

IDEAL SOLUTIONS

As in the crisp version, different interpretations of ideal solutions exist. They can be classified into three main categories (Ploskas & Papathanasiou, 2019):

- Absolute values: $\tilde{v}_j^* = (1, 1, 1)$ and $\tilde{v}_j^- = (0, 0, 0)$ (Chen, 2000)
- Max-min values: $\tilde{v}_j^* = \max_i \tilde{v}_{ij}$ and $\tilde{v}_j^* = \min_i \tilde{v}_{ij}$ (Chen, Lin, & Huang, 2006)
- Fixed values, as in the crisp version (Cables, Lamata, & Verdegay, 2016; García, 2012)

The most commonly used ones are absolute and max-min values.

AGGREGATION

Two main simple aggregations of individual assessments exist in the literature. Let $\tilde{x}_{ijk}(x_{ijk}^1, x_{ijk}^2, x_{ijk}^3, x_{ijk}^4)$ be the kth decision maker's assessment of the ith alternative according to the jth criterion. The group assessment $\tilde{x}_{ij}(x_{ij}^1, x_{ij}^2, x_{ij}^3, x_{ij}^4)$ of the ith alternative according to the jth criterion is generally computed either by mean values (Chen, 2000),

$$x_{ij}^1 = \sum_{k=1}^K x_{ijk}^1/K, \quad x_{ij}^2 = \sum_{k=1}^K x_{ijk}^2/K, \quad x_{ij}^3 = \sum_{k=1}^K x_{ijk}^3/K, \quad x_{ij}^4 = \sum_{k=1}^K x_{ijk}^4/K.$$

or by minimum, mean, and maximum values (Chen, Lin, & Huang, 2006),

$$x_{ij}^1 = \min_k x_{ijk}^1, \quad x_{ij}^2 = \sum_{k=1}^{K} x_{ijk}^2/K, \quad x_{ij}^3 = \sum_{k=1}^{K} x_{ijk}^3/K, \quad x_{ij}^4 = \max_k x_{ijk}^4.$$

Roghanian, Rahimi, and Ansari (2010) compare what they called first aggregation and last aggregation. By first aggregation, they mean merging the different opinions into a collective one as already described, which means just after obtaining the individual assessments and before constructing the normalized decision matrix. By last aggregation, they mean saving the merge until the last step, which implies recalculating the main body of TOPSIS for each decision maker, including the normalized decision matrix and the weighted normalized decision matrix. They conclude that first aggregation is less cumbersome and more precise when variability among opinions is low, whereas last aggregation is more precise with high variability between opinions, which justifies the consensus-reaching approach used later in this chapter.

NORMALIZATION

Practically all normalization methods used in fuzzy TOPSIS are derived from Equation 6.1 or 6.14 for benefit criteria and Equation 6.2 or 6.15 for cost criteria. However, a deeper knowledge of fuzzy logic and especially fuzzy numbers is necessary, which justifies the existence of Chapter 3.

An interval-valued triangular fuzzy number $\tilde{\tilde{A}}$ (Figure 6.2) is composed of two type 1 fuzzy sets $\tilde{A}^L(a^{1^L}, a^{2^L}, a^{3^L})$ and $\tilde{A}^L(a^{1^U}, a^{2^U}, a^{3^U})$. Note that when $a^{2^U} = a^{2^L}$, the interval-valued triangular fuzzy number can be written simply as a function of five tuples: $[(a^{1^U}, a^{1^L}), a^2, (a^{3^L}, a^{3^U})]$.

This means that the interval-valued triangular fuzzy number must respect $a^{1^U} \leq a^{1^L} \leq a^2 \leq a^{3^L} \leq a^{3^U}$.

Trying to extend the TOPSIS methodology using interval-valued triangular fuzzy numbers, Ashtiani et al. (2009) propose the following normalizations for benefit and cost criteria, respectively:

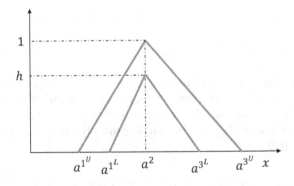

FIGURE 6.2 Interval-Valued Triangular Fuzzy Number.

Fuzzy TOPSIS

$$\tilde{r}_{ij} = \left[\left(x_{ij}^{1^U}/x_j^*, x_{ij}^{1^L}/x_j^* \right), x_{ij}^2/x_j^*, \left(x_{ij}^{3^L}/x_j^*, x_{ij}^{3^U}/x_j^* \right) \right], \tag{6.22}$$

$$\tilde{r}_{ij} = \left[\left(x_j^-/x_{ij}^{1^L}, x_j^-/x_{ij}^{1^U} \right), x_j^-/x_{ij}^2, \left(x_j^-/x_{ij}^{3^U}, x_j^-/x_{ij}^{3^L} \right) \right], \tag{6.23}$$

where $x_j^* = \max_i x_{ij}$ and $x_j^- = \min_i x_{ij}$.

While Equation 6.22 produces a correct interval-valued triangular fuzzy number, the inequalities are not respected with Equation 6.23: it is evident that $x_j^-/x_{ij}^{1^U} \geq x_j^-/x_{ij}^2 \geq x_j^-/x_{ij}^{3^U}$. The authors did not notice this because their example does not contain any cost criteria. Corrections were proposed by Mokhtarian (2015).

DISTANCES

Let $\tilde{A}(a^1, a^2, a^3)$ and $\tilde{B}(b^1, b^2, b^3)$ be two triangular fuzzy numbers. Several distance formulas have been used in the fuzzy TOPSIS algorithm, including the following (Ploskas & Papathanasiou, 2019):

- Chen (2000):

$$d(\tilde{A}, \tilde{B}) = \sqrt{(1/3) * [(a^1 - b^1)^2 + (a^2 - b^2)^2 + (a^3 - b^3)^2]} \tag{6.24}$$

- Jin (2003):

$$d(\tilde{A}, \tilde{B}) = \sqrt{(1/6) * [\sum_{q=1}^{3} (a^q - b^q)^2 + (a^2 - b^2)^2 + \sum_{q=1}^{2} (a^q - b^q) * (a^{q+1} - b^{q+1})]} \tag{6.25}$$

- Chen (1996):

$$d(\tilde{A}, \tilde{B}) = \sum_{q=1}^{3} |a^q - b^q|/3 \tag{6.26}$$

- Chen & Hsieh (1999):

$$d(\tilde{A}, \tilde{B}) = |(a^1 + 4 * a^2 + a^3) - (b^1 + 4 * b^2 + b^3)|/6 \tag{6.27}$$

COMBINATIONS

It is true that some combinations of distance, normalization, ideal solutions, and aggregation are less sensitive to rank reversal than others; however, the choice of combination must in the first place fit the case under consideration. Kahraman et al. (2008) have compared fuzzy TOPSIS methods, as detailed in Table 6.13.

TABLE 6.13
A Comparison of Fuzzy TOPSIS Methods

Reference	Attribute Weights	Fuzzy-Number Type	Ranking Method	Normalization
Chen and Hwang (1992)	Fuzzy numbers	Trapezoidal	Generalized mean (Lee & Li, 1988)	Linear
Liang (1999)	Fuzzy numbers	Trapezoidal	Ranking with maximizing and minimizing set (Chen, 1985)	Linear
Chen (2000)	Fuzzy numbers	Triangular	Vertex (Chen, 2000)	Linear
Chu (2002)	Fuzzy numbers	Triangular	Total integral value with $\alpha = 0.5$ (Liou & Wang, 1992)	Modified Manhattan distance
Tsaur, Chang, and Yen (2002)	Crisp numbers	Triangular	Center of area (Zhao & Govind, 1991)	Vector
Zhang and Lu (2003)	Crisp numbers	Triangular	Vertex mode (Chen, 2000)	Manhattan distance
Chu and Lin (2003)	Fuzzy numbers	Triangular	Mean of removals (Kaufmann & Gupta, 1988)	Linear
Cha and Jung (2003)	Crisp numbers	Triangular	Fuzzy distance operator (Cha & Jung, 2003)	Linear
Yang and Hung (2007)	Fuzzy numbers	Triangular	Vertex (Chen, 2000)	Normalized fuzzy linguistic ratings
Wang and Elhag (2006)	Fuzzy numbers	Triangular	Vertex (Chen, 2000)	Linear
Jahanshahloo, Lotfi and Izadikhah (2006)	Crisp numbers	Interval data	New normalization and ranking (Jahanshahloo, Lotfi, & Izadikhah, 2006)	

OTHER LIMITATIONS

In addition to rank reversal (discussed at length in the previous chapter), another pitfall in TOPSIS is that no consistency or validity checks are incorporated (Madi, Garibaldi, & Wagner, 2016). Suppose an alternative is evaluated as very good according to a specific criterion by the first decision maker and very bad by the second. Obviously the two assessments are not consistent with each other (El Alaoui & El Yassini, 2020). This problem is more persistent in large groups (Wu, Liu, & Liu, 2018). Figure 6.3 shows the result of a SCOPUS database search on "TOPSIS" AND ("group" OR "consensus" OR "consensual" OR "coherence" OR "coherent" OR "coherency").

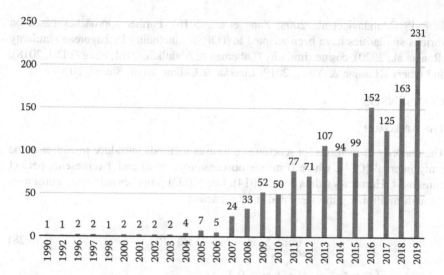

FIGURE 6.3 Group Consensus and TOPSIS in the Scopus Database, 1990–2019.

In a situation involving different people with different backgrounds, it is not sufficient to aggregate their opinions according to a given mathematical function; instead, a consensus must be reached. However, the definition of consensus itself is not consensual. Ideally, consensus would mean unanimity, but such a situation is rarely achievable. In practice, consensus is the aggregation of opinions into one that best represents everyone's (Zhang et al., 2011).

Consensus does not mean a mere majority, either. Assume a situation where 10 people are choosing between two alternatives A_1 and A_2; the votes for, against, and neutral are as follows: (4, 0, 6) for A_1 and (6, 3, 1) for A_2. If we settle the question simply by votes in favor, more people voted for A_2. However, it is far from the consensus choice. The goal is to reach consent, not necessarily agreement (Herrera-Viedma et al., 2014). A simple definition of consensus might be "an agreement of wills without any formal opposition." Hence, the notion of majority is called into question.

Several approaches have been proposed to measure agreement among opinions (Cabrerizo et al., 2017; Ding, Xu, & Liao, 2017; El Alaoui, Ben-azza, & El Yassini, 2019; Lee, 2002; Li, Wang, & Chin, 2019; Liu et al., 2012; Perez et al., 2016; Wu, Chiclana, & Herrera-Viedma, 2015; Xu & Wang, 2013; Xu, Sun, & Wang, 2016; Zhang & Xu, 2015). The notion of similarity (S), particularly in a fuzzy context, plays a prominent role. The more similar two opinions are, the more likely they are to form a consensus (López-Ortega & Castro-Espinoza, 2018). Furthermore, since the second S in TOPSIS stands for similarity, a simple question can be: What is similarity? A widely used approach consists of defining it from its inverse: dissimilarity or distance D (Szmidt, 2014). This approach has been widely used in various applications (El Alaoui, Ben-azza, & El Yassini, 2018, 2019) in what can be called similarity-based distance, or proximity-based measures (Cross & Sudkamp, 2002). The frequently used formula (Santini & Jain, 1997) is $S = 1 - D$. Other formulas exist in the literature, such as $S = 1/(1 + D)$ (Kóczy & Tikk, 2000), and have been applied in

TOPSIS (Mahdavi et al., 2008; Zang et al., 2016). Further approaches based on various similarities have been adapted to TOPSIS, including Pythagorean similarity (Rani et al., 2020), cosine similarity (Otheman & Abdullah, 2014; Peng & Dai, 2018), and others (Hussian & Yang, 2019; Luukka & Collan, 2016; Sharaf, 2018).

CONSENSUS

THE APPROACH

The majority of methods for reaching consensus methods introduce measures in the unity interval [0, 1], where 0 means no consensus at all and 1 represents perfect consensus (Herrera-Viedma et al., 2014). Lee (2002) proposes achieving consensus through minimizing the weighted incoherencies:

$$\min_{M \times IR^4} \sum_{k=1}^{K} g_k^t * (c - S(\tilde{R}_k, \tilde{R})), \qquad (6.28)$$

where $M = \left\{ \begin{array}{l} G = (g_1, g_2, \ldots, g_K), g_k \geq 0, \\ \sum_{k=1}^{K} g_k = 1, \end{array} \right\}$,

t is a positive integer $t > 1$, c is a real number $c > 1$, $\tilde{R}_k(r_k^1, r_k^2, r_k^3, r_k^4)$ are the individual opinions expressed by each DM, $\tilde{R}(r^1, r^2, r^3, r^4)$ is the aggregated consensus, and $S(\tilde{R}_k, \tilde{R})$ is the similarity between the kth decision and the consensus, with

$$S(\tilde{R}_k, \tilde{R}) = 1 - \frac{1}{4u^2}(d(\tilde{R}_k, \tilde{R}))^2 \qquad (6.29)$$

and

$$d(\tilde{R}_k, \tilde{R}) = \left(\sum_{q=1}^{4} (r_k^q - r^q)^2 \right)^{\frac{1}{2}}. \qquad (6.30)$$

While Lee (2002) uses $u = \max_{k,q}(r_k^q) - \min_{k,q}(r_k^q)$, El Alaoui, El Yassini, and Benazza (2019) show that using the whole width of the interval converge in fewer iterations and results in a better value for the optimized function.

The resolution algorithm is as follows:

Algorithm 6.3:

Step 1: Each DM express their opinion using a trapezoidal fuzzy number $\tilde{R}_k$.

Step 2: Fix initial opinion weights $G^{(0)}(g_1^{(0)}, \ldots, g_K^{(0)})$ satisfying $0 \leq g_k^{(0)} \leq 1$ and $\sum_{k=1}^{K} g_k = 1$.

The iterations will be indicated by $l = 0, 1, \ldots$.

Fuzzy TOPSIS

Step 3: Compute

$$\widetilde{\widetilde{R}}^{(l+1)} = \frac{\sum_{k=1}^{K} g_k^{(l)^t} \otimes \widetilde{\widetilde{R}}_k}{\sum_{k=1}^{K} g_k^{(l)^t}}$$

Step 4: Compute

$$g_k^{(l+1)} = \frac{(1/(c - S(\widetilde{\widetilde{R}}_k, \widetilde{\widetilde{R}}^{(l+1)})))^{1/(t-1)}}{\sum_{k=1}^{K} (1/(c - S(\widetilde{\widetilde{R}}_k, \widetilde{\widetilde{R}}^{(l+1)})))^{1/(t-1)}}.$$

Step 5: If $G^{(l+1)} - G^{(l)} \leq \varepsilon$, stop; otherwise, set $l = l + 1$ and go to step 3. □

Example

In an example (Table 6.14) treated by multiple authors (El Alaoui 2018; El Alaoui, El Yassini, & Ben-azza 2019; Tan, Cai, & Qi 2010), three closed-loop agricultural supply-chain alternatives A_1, A_2, and A_3 are evaluated according to four criteria—C_1: reliability; C_2: effectiveness; C_3: cost; and C_4: asset management.

It is true that all methods result in the same ranking: $A_3 \geq A_1 \geq A_2$. However, El Alaoui (2018) and El Alaoui, El Yassini, and Ben-azza (2019) have shown experimentally that adopting a consensus-reaching process produces results that are more adequate to the definition of TOPSIS. While A_3 is both the closest to the positive ideal solution and the furthest from the negative ideal solution in all methods, only those of El Alaoui (2018) and El Alaoui et al. (2019) achieve $D_2^* > D_1^*$ and $D_2^- < D_1^-$ (Table 6.15).

PROPOSED FUZZY TOPSIS ALGORITHM

THE APPROACH

The majority of fuzzy TOPSIS approaches deal exclusively with qualitative criteria. Here an algorithm is proposed taking in consideration both quantitative and qualitative criteria.

TABLE 6.14
Fuzzy TOPSIS Example with Consensus Measure

Decision Maker	C_1			C_2			C_3			C_4		
	A_1	A_2	A_3	A_1	A_2	A_3	A_1	A_2	A_3	A_1	A_2	A_3
DM_1	MH	F	MH	VH	MH	H	F	MH	VH	F	MH	VH
DM_2	F	F	MH	H	F	MH	F	F	H	F	MH	MH
DM_3	MH	MH	H	VH	MH	VH	MH	H	VH	MH	H	VH

TABLE 6.15
Distances to Ideal Solutions for Fuzzy TOPSIS with Consensus Measure (El Alaoui et al., 2019)

Alternatives	D_i^*	D_i^-	CC_i	rank
A_1	1.3931	2.7589	0.6645	2
A_2	1.4587	2.7057	0.6497	3
A_3	0.6679	3.4631	0.8383	1

Algorithm 6.4:

Step 1: Each decision maker assesses each alternative according to qualitative criteria using a linguistic variable (Table 6.6).

Step 2: Each decision maker assesses each criterion's importance using a linguistic variable (Table 6.6).

Step 3: Compute the group assessments for criterion weights and alternatives according to qualitative criteria using Algorithm 6.3.

Step 4: Construct the normalized decision matrix using Equation 6.14 for benefit criteria and Equation 6.15 for cost criteria.

Step 5: Construct the weighted normalized decision matrix (Equation 6.3).

Step 6: Compute distances to positive (Equation 6.19) and negative (Equation 6.20) ideal solutions.

Step 7: Compute closeness coefficients (Equation 6.21).

Step 8: Rank alternatives according to closeness coefficients, from highest to lowest.

EXAMPLE

In the example treated by Zhao, Zhao, and Guo (2017) and El Alaoui (2020), six alternatives of eco-industrial parks are evaluated by four decision makers according to 26 criteria—nine qualitative and 15 quantitative.

Step 1: Each decision maker assesses each alternative according to qualitative criteria using a linguistic variable (Table 6.16).

Step 2: Each decision maker assesses each criterion's importance using a linguistic variable.

Table 6.17 details weights assessments by each decision maker in addition to criterion type and nature.

TABLE 6.16
Decision Makers Individual Assessments for Each Alternative According to Qualitative Criteria by Linguistic Variables

Decision Maker	Alternative	Qualitative Criterion								
		C_5	C_{11}	C_{18}	C_{19}	C_{20}	C_{22}	C_{23}	C_{24}	C_{25}
DM_1	A_1	G	G	G	G	G	G	G	G	G
	A_2	F	G	F	G	G	G	G	F	G
	A_3	G	F	VG	F	F	G	F	F	F
	A_4	VG	F	G	F	F	G	F	F	G
	A_5	G	G	P	P	G	F	F	F	F
	A_6	F	P	F	F	VP	F	F	F	P
DM_2	A_1	VG	G	G	G	G	G	G	G	G
	A_2	G	F	F	F	VG	G	F	F	F
	A_3	G	F	G	G	F	F	G	F	F
	A_4	VG	F	F	G	F	F	F	F	G
	A_5	G	G	VP	P	F	P	F	F	F
	A_6	F	P	P	F	P	P	P	P	F
DM_3	A_1	G	G	G	VG	G	G	G	G	G
	A_2	G	G	F	F	G	F	F	F	G
	A_3	G	F	G	F	P	F	F	F	F
	A_4	VG	F	F	G	G	G	G	F	F
	A_5	G	G	VP	F	G	P	G	F	F
	A_6	F	P	P	P	P	P	F	F	P
DM_4	A_1	G	F	G	G	G	G	G	G	VG
	A_2	G	F	F	F	G	F	F	F	G
	A_3	G	F	G	G	F	F	F	F	F
	A_4	G	F	F	G	F	G	F	F	F
	A_5	G	G	VP	P	F	F	F	F	F
	A_6	F	F	P	P	P	P	F	F	F

Step 3: Compute the group assessments for criterion weights and alternatives according to qualitative criteria using Algorithm 6.3.

The consensual assessments for criterion weights are detailed in Table 6.17, and those for qualitative criteria are presented in Table 6.18.

Step 4: Construct the normalized decision matrix.

The normalized alternative assessments according to quantitative criteria are shown in Table 6.19.

The normalized alternative assessments in accordance with qualitative criteria are given in Table 6.20.

Step 5: Construct the weighted normalized decision matrix (Table 6.21).

Step 6: Compute distances to positive and negative ideal solutions (Table 6.22).

TABLE 6.17
Weight Assessment, Criterion Type, Nature, and Consensual Assessment

Criterion	Type	Nature	DM_1	DM_2	DM_3	DM_4	Consensual Assessment
C_1	Quantitative	Max	H	H	M	M	(0.5, 0.7, 0.85)
C_2	Quantitative	Max	H	M	H	H	(0.62, 0.82, 0.94)
C_3	Quantitative	Max	M	M	H	M	(0.38, 0.58, 0.76)
C_4	Quantitative	Max	L	L	L	VL	(0, 0.08, 0.25)
C_5	qualitative	Max	M	H	M	M	(0.38, 0.58, 0.76)
C_6	Quantitative	Min	H	H	H	M	(0.62, 0.82, 0.94)
C_7	Quantitative	Min	M	H	H	M	(0.5, 0.7, 0.85)
C_8	Quantitative	Max	H	H	H	H	(0.7, 0.9, 1)
C_9	Quantitative	Max	M	M	H	H	(0.5, 0.7, 0.85)
C_{10}	Quantitative	Max	H	H	H	H	(0.7, 0.9, 1)
C_{11}	qualitative	Max	M	M	H	H	(0.5, 0.7, 0.85)
C_{12}	Quantitative	Min	H	H	H	M	(0.62, 0.82, 0.94)
C_{13}	Quantitative	Min	H	M	H	M	(0.5, 0.7, 0.85)
C_{14}	Quantitative	Min	H	H	H	M	(0.62, 0.82, 0.94)
C_{15}	Quantitative	Max	H	H	H	H	(0.7, 0.9, 1)
C_{16}	Quantitative	Max	H	H	H	H	(0.7, 0.9, 1)
C_{17}	Quantitative	Max	H	H	H	H	(0.7, 0.9, 1)
C_{18}	qualitative	Max	H	H	H	H	(0.7, 0.9, 1)
C_{19}	qualitative	Max	M	M	H	M	(0.38, 0.58, 0.76)
C_{20}	qualitative	Max	M	H	H	M	(0.5, 0.7, 0.85)
C_{21}	Quantitative	Max	M	M	H	H	(0.5, 0.7, 0.85)
C_{22}	qualitative	Max	M	M	H	H	(0.5, 0.7, 0.85)
C_{23}	qualitative	Max	M	H	M	M	(0.38, 0.58, 0.76)
C_{24}	qualitative	Max	M	H	M	M	(0.38, 0.58, 0.76)
C_{25}	qualitative	Max	M	H	H	M	(0.5, 0.7, 0.85)
C_{26}	Quantitative	Max	M	M	M	H	(0.38, 0.58, 0.76)

Step 7: Compute closeness coefficients (Table 6.22).

Step 8: Rank alternatives according to closeness coefficients, from highest to lowest (Table 6.22).

While the ranking obtained is similar to the one proposed by Zhao, Zhao, and Guo (2017), they opt for early defuzzification with simple additive weighting, which can produce information loss (Madi, Garibaldi, & Wagner, 2017; Sang & Liu, 2016). In the proposed method, in addition to the consensus achieved, no defuzzification is required, preserving all information. The ranking also perfectly fulfills the definition of TOPSIS, since each alternative outperforms the next one, being closer to the positive ideal solution and further from the negative one: $D_1^* < D_2^* < D_4^* < D_3^* < D_5^* < D_6^*$ and $D_1^- > D_2^- > D_4^- > D_3^- > D_5^- > D_6^-$.

TABLE 6.18
Consensual Alternatives Assessment According to Qualitative Criteria

Qualitative Criterion	Alternative					
	A_1	A_2	A_3	A_4	A_5	A_6
C_5	(7.49, 9.25, 10)	(6.16, 8.16, 9.37)	(7, 9, 10)	(8.51, 9.75, 10)	(7, 9, 10)	(3, 5, 7)
C_{11}	(6.16, 8.16, 9.37)	(5, 7, 8.5)	(3, 5, 7)	(3, 5, 7)	(7, 9, 10)	(0.63, 1.84, 3.84)
C_{18}	(7, 9, 10)	(3, 5, 7)	(7.49, 9.25, 10)	(3.84, 5.84, 7.63)	(0, 0.25, 1.49)	(0.63, 1.84, 3.84)
C_{19}	(7.49, 9.25, 10)	(3.84, 5.84, 7.63)	(5, 7, 8.5)	(6.16, 8.16, 9.37)	(0.63, 1.84, 3.84)	(1.5, 3, 5)
C_{20}	(7, 9, 10)	(7.49, 9.25, 10)	(2.37, 4.16, 6.16)	(3.84, 5.84, 7.63)	(5, 7, 8.5)	(0, 0.75, 2.51)
C_{22}	(7, 9, 10)	(5, 7, 8.5)	(3.84, 5.84, 7.63)	(6.16, 8.16, 9.37)	(1.5, 3, 5)	(0.63, 1.84, 3.84)
C_{23}	(7, 9, 10)	(3.84, 5.84, 7.63)	(3.84, 5.84, 7.63)	(3.84, 5.84, 7.63)	(3.84, 5.84, 7.63)	(2.37, 4.16, 6.16)
C_{24}	(7, 9, 10)	(3, 5, 7)	(3, 5, 7)	(3, 5, 7)	(3, 5, 7)	(2.37, 4.16, 6.16)
C_{25}	(7.49, 9.25, 10)	(6.16, 8.16, 9.37)	(3, 5, 7)	(5, 7, 8.5)	(3, 5, 7)	(1.5, 3, 5)

INTERVAL TYPE 2 FUZZY TOPSIS

THE APPROACH

While a large number of interval type 2 fuzzy TOPSIS extensions and applications use interval type 2 fuzzy numbers (Büyüközkan, Parlak, & Tolga, 2014, 2016; Cengiz Toklu, 2018; Ilieva, 2016; Liao, 2015; Mei & Xie, 2019; Nehi & Keikha, 2016; Sang & Liu, 2016; Sharaf, 2019; Wang et al., 2019; Wu, Liu, & Liu, 2018; Yang, Liu, & Liu, 2020), fewer consider perfect interval type 2 fuzzy numbers (Chen & Hong, 2014; Chen & Lee, 2010; Dymova, Sevastjanov, & Tikhonenko, 2015; Shyi-Ming & Hong, 2014; Yue et al., 2019; Zamri & Abdullah, 2014)—or

TABLE 6.19
Normalized Alternative Assessments for Quantitative Criteria

Quantitative Criterion	Alternative					
	A_1	A_2	A_3	A_4	A_5	A_6
C_1	0.15949	0.07407	0.24958	0.24394	0.21496	0.05797
C_2	0.21742	0.21212	0.19697	0.17424	0.18939	0.00985
C_3	0.20549	0.24631	0.19704	0.21675	0.09923	0.03519
C_4	0.18479	0.15529	0.13976	0.28061	0.14162	0.09793
C_6	0.12228	0.20380	0.12738	0.24456	0.14912	0.15285
C_7	0.14980	0.17174	0.33371	0.15898	0.15131	0.03446
C_8	0.18292	0.16781	0.18909	0.17984	0.13647	0.14387
C_9	0.16393	0.11241	0.07728	0.19672	0.09836	0.35129
C_{10}	0.17978	0.17127	0.18945	0.16180	0.13145	0.16625
C_{12}	0.19064	0.19064	0.11255	0.31139	0.12292	0.07186
C_{13}	0.17669	0.22087	0.29449	0.17669	0.12621	0.00505
C_{14}	0.17707	0.27826	0.14428	0.16232	0.19478	0.04328
C_{15}	0.17584	0.17407	0.17762	0.17052	0.16874	0.13321
C_{16}	0.17331	0.17331	0.16464	0.16811	0.15598	0.16464
C_{17}	0.17422	0.16899	0.17422	0.17247	0.17073	0.13937
C_{21}	0.19280	0.18123	0.17138	0.12725	0.14996	0.17738
C_{26}	0.36611	0.34728	0.06485	0.12343	0.07950	0.01883

even both (Dymova, Sevastjanov, & Tikhonenko, 2016). However, as mentioned in Chapter 3, only perfect type 2 fuzzy numbers, in which the lower and upper bounds are both type 1 fuzzy numbers, deserve to be called interval type 2 fuzzy numbers (Mendel, 2017). Hence, to adapt to perfect interval type 2 fuzzy sets, the similarity (Equation 6.29) and distance (Equation 6.30) from Algorithm 6.3 become

$$S(\tilde{R}_k, \tilde{R}) = 1 - \frac{1}{8}(d(\tilde{R}_k, \tilde{R}))^2 \qquad (6.31)$$

and

$$d(\tilde{R}_k, \tilde{R}) = \left(\sum_{q=1}^{4} (r_k^{qL} - r^{qL})^2 + \sum_{q=1}^{4} (r_k^{qU} - r^{qU})^2 \right)^{\frac{1}{2}}, \qquad (6.32)$$

where $\tilde{R}_k$ ($[r_k^{1L}, r_k^{2L}, r_k^{3L}, r_k^{4L}]$, $[r_k^{1U}, r_k^{2U}, r_k^{3U}, r_k^{4U}]$) is the kth individual opinion and $\tilde{R}([r^{1L}, r^{2L}, r^{3L}, r^{4L}], [r^{1U}, r^{2U}, r^{3U}, r^{4U}])$ is the aggregated consensus.

Thus, the resolution algorithm becomes the following:

Fuzzy TOPSIS

TABLE 6.20
Normalized Alternative Assessments for Qualitative Criteria

Qualitative Criterion	Alternative					
	A_1	A_2	A_3	A_4	A_5	A_6
C_5	(0.75, 0.92, 1)	(0.62, 0.82, 0.94)	(0.7, 0.9, 1)	(0.85, 0.98, 1)	(0.7, 0.9, 1)	(0.3, 0.5, 0.7)
C_{11}	(0.62, 0.82, 0.94)	(0.5, 0.7, 0.85)	(0.3, 0.5, 0.7)	(0.3, 0.5, 0.7)	(0.7, 0.9, 1)	(0.06, 0.18, 0.38)
C_{18}	(0.7, 0.9, 1)	(0.3, 0.5, 0.7)	(0.75, 0.92, 1)	(0.38, 0.58, 0.76)	(0, 0.02, 0.15)	(0.06, 0.18, 0.38)
C_{19}	(0.75, 0.92, 1)	(0.38, 0.58, 0.76)	(0.5, 0.7, 0.85)	(0.62, 0.82, 0.94)	(0.06, 0.18, 0.38)	(0.15, 0.3, 0.5)
C_{20}	(0.7, 0.9, 1)	(0.75, 0.92, 1)	(0.24, 0.42, 0.62)	(0.38, 0.58, 0.76)	(0.5, 0.7, 0.85)	(0, 0.08, 0.25)
C_{22}	(0.7, 0.9, 1)	(0.5, 0.7, 0.85)	(0.38, 0.58, 0.76)	(0.62, 0.82, 0.94)	(0.15, 0.3, 0.5)	(0.06, 0.18, 0.38)
C_{23}	(0.7, 0.9, 1)	(0.38, 0.58, 0.76)	(0.38, 0.58, 0.76)	(0.38, 0.58, 0.76)	(0.38, 0.58, 0.76)	(0.24, 0.42, 0.62)
C_{24}	(0.7, 0.9, 1)	(0.3, 0.5, 0.7)	(0.3, 0.5, 0.7)	(0.3, 0.5, 0.7)	(0.3, 0.5, 0.7)	(0.24, 0.42, 0.62)
C_{25}	(0.75, 0.92, 1)	(0.62, 0.82, 0.94)	(0.3, 0.5, 0.7)	(0.5, 0.7, 0.85)	(0.3, 0.5, 0.7)	(0.15, 0.3, 0.5)

Algorithm 6.5:

Let K decision makers ($1 \leq k \leq K$) assess m alternatives ($1 \leq i \leq m$) according to n criteria ($1 \leq j \leq n$).

Step 1: Each decision maker assesses each alternative using a linguistic variable $\tilde{x}_{ijk}$.

TABLE 6.21
Weighted Normalized Decision Matrix

Criterion	Alternative					
	A_1	A_2	A_3	A_4	A_5	A_6
C_1	(0.08, 0.11, 0.14)	(0.04, 0.05, 0.06)	(0.12, 0.17, 0.21)	(0.12, 0.17, 0.21)	(0.11, 0.15, 0.18)	(0.03, 0.04, 0.05)
C_2	(0.13, 0.18, 0.2)	(0.13, 0.17, 0.2)	(0.12, 0.16, 0.18)	(0.11, 0.14, 0.16)	(0.12, 0.15, 0.18)	(0.01, 0.01, 0.01)
C_3	(0.08, 0.12, 0.16)	(0.09, 0.14, 0.19)	(0.08, 0.12, 0.15)	(0.08, 0.13, 0.17)	(0.04, 0.06, 0.08)	(0.01, 0.02, 0.03)
C_4	(0, 0.01, 0.05)	(0, 0.01, 0.04)	(0, 0.01, 0.04)	(0, 0.02, 0.7)	(0, 0.01, 0.04)	(0, 0.01, 0.02)
C_5	(0.29, 0.54, 0.76)	(0.24, 0.48, 0.71)	(0.27, 0.53, 0.76)	(0.33, 0.57, 0.76)	(0.27, 0.53, 0.76)	(0.12, 0.29, 0.53)
C_6	(0.08, 0.1, 0.11)	(0.13, 0.17, 0.19)	(0.08, 0.1, 0.12)	(0.15, 0.2, 0.23)	(0.09, 0.12, 0.14)	(0.09, 0.12, 0.14)
C_7	(0.07, 0.1, 0.13)	(0.09, 0.12, 0.15)	(0.17, 0.23, 0.25)	(0.08, 0.11, 0.14)	(0.08, 0.11, 0.13)	(0.02, 0.02, 0.03)
C_8	(0.13, 0.16, 0.18)	(0.12, 0.15, 0.17)	(0.13, 0.17, 0.19)	(0.13, 0.16, 0.18)	(0.1, 0.12, 0.14)	(0.1, 0.13, 0.14)
C_9	(0.08, 0.11, 0.14)	(0.06, 0.08, 0.1)	(0.04, 0.05, 0.07)	(0.1, 0.14, 0.17)	(0.05, 0.07, 0.08)	(0.18, 0.25, 0.3)
C_{10}	(0.13, 0.16, 0.18)	(0.12, 0.15, 0.17)	(0.13, 0.17, 0.19)	(0.11, 0.15, 0.16)	(0.09, 0.12, 0.13)	(0.12, 0.15, 0.17)
C_{11}	(0.31, 0.57, 0.8)	(0.25, 0.49, 0.72)	(0.15, 0.35, 0.6)	(0.15, 0.35, 0.6)	(0.35, 0.63, 0.85)	(0.03, 0.13, 0.33)
C_{12}	(0.12, 0.16, 0.18)	(0.12, 0.16, 0.18)	(0.07, 0.09, 0.11)	(0.19, 0.25, 0.29)	(0.08, 0.1, 0.12)	(0.04, 0.06, 0.07)
C_{13}	(0.09, 0.12, 0.15)	(0.11, 0.15, 0.19)	(0.15, 0.21, 0.25)	(0.09, 0.12, 0.15)	(0.06, 0.09, 0.11)	(0.00, 0.00, 0.00)

(*Continued*)

TABLE 6.21 (Continued)

Criterion	Alternative					
	A_1	A_2	A_3	A_4	A_5	A_6
C_{14}	(0.11, 0.14, 0.17)	(0.17, 0.23, 0.26)	(0.09, 0.12, 0.15)	(0.1, 0.13, 0.15)	(0.12, 0.16, 0.18)	(0.03, 0.04, 0.04)
C_{15}	(0.12, 0.16, 0.18)	(0.12, 0.16, 0.17)	(0.12, 0.16, 0.18)	(0.12, 0.15, 0.17)	(0.12, 0.15, 0.17)	(0.12, 0.15, 0.16)
C_{16}	(0.12, 0.16, 0.17)	(0.12, 0.16, 0.17)	(0.12, 0.15, 0.16)	(0.12, 0.15, 0.17)	(0.11, 0.14, 0.16)	(0.1, 0.13, 0.14)
C_{17}	(0.12, 0.16, 0.17)	(0.12, 0.15, 0.17)	(0.12, 0.16, 0.17)	(0.12, 0.16, 0.17)	(0.12, 0.15, 0.17)	(0.04, 0.17, 0.38)
C_{18}	(0.49, 0.81, 1)	(0.21, 0.45, 0.7)	(0.52, 0.83, 1)	(0.27, 0.53, 0.76)	(0, 0.02, 0.15)	(0.06, 0.18, 0.38)
C_{19}	(0.29, 0.54, 0.76)	(0.15, 0.34, 0.58)	(0.19, 0.41, 0.65)	(0.24, 0.48, 0.71)	(0.02, 0.11, 0.29)	(0, 0.05, 0.21)
C_{20}	(0.35, 0.63, 0.85)	(0.37, 0.65, 0.85)	(0.12, 0.29, 0.52)	(0.19, 0.41, 0.65)	(0.25, 0.49, 0.72)	(0.09, 0.12, 0.15)
C_{21}	(0.1, 0.13, 0.16)	(0.09, 0.13, 0.15)	(0.09, 0.12, 0.15)	(0.06, 0.09, 0.11)	(0.07, 0.1, 0.13)	(0.03, 0.13, 0.33)
C_{22}	(0.35, 0.63, 0.85)	(0.25, 0.49, 0.72)	(0.19, 0.41, 0.65)	(0.31, 0.57, 0.8)	(0.08, 0.21, 0.43)	(0.09, 0.24, 0.47)
C_{23}	(0.27, 0.53, 0.76)	(0.15, 0.34, 0.58)	(0.15, 0.34, 0.58)	(0.15, 0.34, 0.58)	(0.15, 0.34, 0.58)	(0.09, 0.24, 0.47)
C_{24}	(0.27, 0.53, 0.76)	(0.12, 0.29, 0.53)	(0.12, 0.29, 0.53)	(0.12, 0.29, 0.53)	(0.12, 0.29, 0.53)	(0.09, 0.24, 0.47)
C_{25}	(0.37, 0.65, 0.85)	(0.31, 0.57, 0.8)	(0.15, 0.35, 0.59)	(0.25, 0.49, 0.72)	(0.15, 0.35, 0.59)	(0.08, 0.21, 0.43)
C_{26}	(0.14, 0.21, 0.28)	(0.13, 0.2, 0.27)	(0.02, 0.04, 0.05)	(0.05, 0.07, 0.09)	(0.03, 0.05, 0.06)	(0.01, 0.01, 0.01)

TABLE 6.22
Distances to Ideal Solutions, Closeness Coefficients, and Final Ranking

Alternative	D_i^*	D_i^-	CC_i	Rank
A_1	18.9	7.87	0.29	1
A_2	19.9	6.83	0.26	2
A_3	20.3	6.42	0.24	4
A_4	20	6.73	0.25	3
A_5	21.3	5.27	0.2	5
A_6	22.9	3.64	0.14	6

Step 2: Each decision maker assesses each criterion's importance using a linguistic variable $\tilde{w}_{jk}$.

Step 3: Using the individual alternative assessments $\tilde{x}_{ijk}([x_{ijk}^{1^L}, x_{ijk}^{2^L}, x_{ijk}^{3^L}, x_{ijk}^{4^L}], [x_{ijk}^{1^U}, x_{ijk}^{2^U}, x_{ijk}^{3^U}, x_{ijk}^{4^U}])$ and criterion weights $\tilde{w}_{jk}([w_{jk}^{1^L}, w_{jk}^{2^L}, w_{jk}^{3^L}, w_{jk}^{4^L}], [w_{jk}^{1^U}, w_{jk}^{2^U}, w_{jk}^{3^U}, w_{jk}^{4^U}])$, compute the collective alternative assessments $\tilde{x}_{ij}$ and criterion weights $\tilde{w}_j$ using Algorithm 6.3 and Equations 6.31 and 6.32.

Step 4: Construct the normalized decision matrix $\tilde{R} = [\tilde{r}_{ij}]_{m \times n}$.

For benefit criteria,

$$\tilde{r}_{ij} = \left(\frac{x_{ij}^1}{x_{ij}^*}, \frac{x_{ij}^2}{x_{ij}^*}, \frac{x_{ij}^3}{x_{ij}^*}, \frac{x_{ij}^4}{x_{ij}^*} \right) \quad (6.33)$$

with $x_{ij}^* = \max_i x_{ij}^4$.

For cost criteria,

$$\tilde{r}_{ij} = \left(\frac{x_{ij}^-}{x_{ij}^1}, \frac{x_{ij}^-}{x_{ij}^2}, \frac{x_{ij}^-}{x_{ij}^3}, \frac{x_{ij}^-}{x_{ij}^4} \right) \quad (6.34)$$

with $x_{ij}^- = \min_i x_{ij}^1$.

Step 5: Construct the weighted normalized decision matrix $\tilde{V} = [\tilde{v}_{ij}]_{m \times n}$.

$$\tilde{v}_{ij} = \tilde{x}_{ij} \otimes \tilde{w}_j \quad (6.35)$$

Fuzzy TOPSIS

such that $v_{ij}^{q^L} = x_{ij}^{q^L} \times w_j^{q^L}$ and $v_{ij}^{q^U} = x_{ij}^{q^U} \times w_j^{q^U}$, with $1 \leq q \leq 4$.

Step 6: Determine the ideal solutions.

The positive ideal solution is

$$A^* = \{v_1^*, \ldots, v_n^*\}, \tag{6.36}$$

with $\tilde{v}_j^* = ([1, 1, 1, 1], [1, 1, 1, 1])$ for benefit criteria and $\tilde{v}_j^* = ([0, 0, 0, 0], [0, 0, 0, 0])$ for cost criteria.

The negative ideal solution is

$$A^* = \{v_1^-, \ldots, v_n^-\}, \tag{6.37}$$

with $\tilde{v}_j^- = ([0, 0, 0, 0], [0, 0, 0, 0])$ for benefit criteria and $\tilde{v}_j^- = ([1, 1, 1, 1], [1, 1, 1, 1])$ for cost criteria.

Step 7: Compute the distance between each alternative and the ideal solutions.

The distance to the positive ideal solution is

$$D_i^* = \sum_{j=1}^n d(\tilde{v}_{ij}, \tilde{v}_j^*), \ i = 1, \ldots, m, \tag{6.38}$$

with $d(\tilde{v}_{ij}, \tilde{v}_j^*) = \sqrt{\left(\frac{1}{8}\right) \times \sum_{q=1}^4 (1 - v_{ij}^{q^L})^2 + (1 - v_{ij}^{q^U})^2}$.

The distance to the negative ideal solution is

$$D_i^- = \sum_{j=1}^n d(\tilde{v}_{ij}, \tilde{v}_j^-), \ i = 1, \ldots m, \tag{6.39}$$

with $d(\tilde{v}_{ij}, \tilde{v}_j^-) = \sqrt{\left(\frac{1}{8}\right) * \sum_{q=1}^4 (v_{ij}^{q^L})^2 + (v_{ij}^{q^U})^2}$.

Step 8: Calculate the relative closeness to the ideal solutions.

$$CC_i = \frac{D_i^-}{D_i^- + D_i^*}, \ i = 1, \ldots, m \tag{6.40}$$

Step 9: Rank alternatives according to CC_i values, from highest to lowest. □

Example

In this example from Chen and Hong (2014), three decision makers (DM_1, DM_2, and DM_3) are to select a candidate from among four alternatives (A_1, A_2, A_3, and A_4) according to two criteria (C_1: emotional steadiness; C_2: oral communication skill).

Step 1: Each decision maker assesses each alternative using a linguistic variable $\tilde{x}_{ijk}$ (Table 6.23).

Step 2: Each decision maker assesses each criterion's importance using a linguistic variable $\tilde{w}_{jk}$ (Table 6.24).

TABLE 6.23
Individual and Collective Alternative Assessment

Criterion	Alternative	Decision Maker			Consensual Evaluation
		DM_1	DM_2	DM_3	
C_1	A_1	VP	P	VP	([0, 0.2676, 0.2676 1.5353], [0, 0.2676, 0.2676 1.5353])
	A_2	MP	MP	P	([0.8192, 2.6385, 2.6385 4.6385], [0.8192, 2.6385, 2.6385 4.6385])
	A_3	P	MP	P	([0.1808, 1.3615, 1.3615, 3.3615], [0.1808, 1.3615, 1.3615, 3.3615])
	A_4	MP	MP	P	([0.8192, 2.6385, 2.6385, 4.6385], [0.8192, 2.6385, 2.6385, 4.6385])
C_2	A_1	M	M	MG	([3.2994, 5.2994, 5.2994, 7.2994], [3.2994, 5.2994, 5.2994, 7.2994])
	A_2	M	MG	M	([3.2994, 5.2994, 5.2994, 7.2994], [3.2994, 5.2994, 5.2994, 7.2994])
	A_3	MG	MG	G	([5.3616, 7.3616, 7.3616, 9.1808], [5.3616, 7.3616, 7.3616, 9.1808])
	A_4	G	G	G	([7, 9, 9, 10], [7, 9, 9, 10])

TABLE 6.24
Individual and Collective Criterion Assessments

Criterion	Decision Maker			Consensual Evaluation
	DM_1	DM_2	DM_3	
C_1	MH	H	MH	([0.5664, 0.7664, 0.7664, 0.9332], [0.5664, 0.7664, 0.7664, 0.9332])
C_2	MH	H	MH	([0.5664, 0.7664, 0.7664, 0.9332], [0.5664, 0.7664, 0.7664, 0.9332])

TABLE 6.25
Linguistic Assessment for Criteria and Alternatives

Linguistic Variable for Criteria	Corresponding Interval-Valued Type 2 Fuzzy Number	Linguistic Variable for Alternatives	Corresponding Interval-Valued Type 2 Fuzzy Number
Very low (VL)	([0, 0, 0, 0.1], [0, 0, 0, 0.1])	Very poor (VP)	([0, 0, 0, 1], [0, 0, 0, 1])
Low (L)	([0, 0.1, 0.1, 0.3], [0, 0.1, 0.1, 0.3])	Poor (P)	([0, 1, 1, 3], [0, 1, 1, 3])
Medium low (ML)	([0.1, 0.3, 0.3, 0.5], [0.1, 0.3, 0.3, 0.5])	Medium poor (MP)	([1, 3, 3, 5], [1, 3, 3, 5])
Medium (M)	([0.3, 0.5, 0.5, 0.7], [0.3, 0.5, 0.5, 0.7])	Medium (M)	([3, 5, 5, 7], [3, 5, 5, 7])
Medium high (MH)	([0.5, 0.7, 0.7, 0.9], [0.5, 0.7, 0.7, 0.9])	Medium good (MG)	([5, 7, 7, 9], [5, 7, 7, 9])
High (H)	([0.7, 0.9, 0.9, 1], [0.7, 0.9, 0.9, 1])	Good (G)	([7, 9, 9, 10], [7, 9, 9, 10])
Very high (VH)	([0.9, 1, 1, 1], [0.9, 1, 1, 1])	Very good (VG)	([9, 10, 10, 10], [9, 10, 10, 10])

TABLE 6.26
Normalized Decision Matrix

Criterion	Alternative	Normalized Assessment
C_1	A_1	([0, 0.0268, 0.0268, 0.1535], [0, 0.0268, 0.0268, 0.1535])
	A_2	([0.0819, 0.2638, 0.2638, 0.4638], [0.0819, 0.2638, 0.2638, 0.4638])
	A_3	([0.0181, 0.1362, 0.1362, 0.3362], [0.0181, 0.1362, 0.1362, 0.3362])
	A_4	([0.0819, 0.2638, 0.2638, 0.4638], [[0.0819, 0.2638, 0.2638, 0.4638])
C_2	A_1	([0.3299, 0.5299, 0.5299, 0.7299], [0.3299, 0.5299, 0.5299, 0.7299])
	A_2	([0.3299, 0.5299, 0.5299, 0.7299], [0.3299, 0.5299, 0.5299, 0.7299])
	A_3	([0.5362, 0.7362, 0.7362, 0.9181], [0.5362, 0.7362, 0.7362, 0.9181])
	A_4	([0.7, 0.9, 0.9, 1], [0.7, 0.9, 0.9, 1])

Step 3: Compute the collective alternative assessments (Table 6.23) $\tilde{x}_{ij}$ and criterion weights $\tilde{w}_j$ (Table 6.24) using Algorithm 6.3 and Equations 6.31 and 6.32 (Table 6.25).

Step 4: Construct the normalized decision matrix $\tilde{R} = [\tilde{r}_{ij}]_{m \times n}$ (Table 6.26).

Step 5: Construct the weighted normalized decision matrix $\tilde{V} = [\tilde{v}_{ij}]_{m \times n}$ (Table 6.27).

Step 6: Determine the ideal solutions.

TABLE 6.27
Weighted Normalized Decision Matrix

Criterion	Alternative	Weighted Normalized Assessment
C_1	A_1	([0, 0.0205, 0.0205, 0.1433], [0, 0.0205, 0.0205, 0.1433])
	A_2	([0.0464, 0.2022, 0.2022, 0.4329], [0.0464, 0.2022, 0.2022, 0.4329])
	A_3	([0.0102, 0.1044, 0.1044, 0.3137], [0.0102, 0.1044, 0.1044, 0.3137])
	A_4	([0.0464, 0.2022, 0.2022, 0.4329], [0.0464, 0.2022, 0.2022, 0.4329])
C_2	A_1	([0.1869, 0.4062, 0.4062, 0.6812], [0.1869, 0.4062, 0.4062, 0.6812])
	A_2	([0.1869, 0.4062, 0.4062, 0.6812], [0.1869, 0.4062, 0.4062, 0.6812])
	A_3	([0.3037, 0.5642, 0.5642, 0.8568], [0.3037, 0.5642, 0.5642, 0.8568])
	A_4	([0.3965, 0.6898, 0.6898, 0.9332], [0.3965, 0.6898, 0.6898, 0.9332])

TABLE 6.28
Distances to Ideal Solutions, Closeness Coefficients, and Final Ranking

Alternative	D_i^*	D_i^-	CC_i	Rank
A_1	1.5614	0.5283	0.2528	4
A_2	1.3970	0.7157	0.3387	3
A_3	1.3444	0.7782	0.3666	2
A_4	1.1657	0.9639	0.4526	1

The positive ideal solution is

$$A^* = \{([1, 1, 1, 1], [1, 1, 1, 1]), ([1, 1, 1, 1], [1, 1, 1, 1])\}.$$

The negative ideal solution is

$$A^- = \{([0, 0, 0, 0], [0, 0, 0, 0]), ([0, 0, 0, 0], [0, 0, 0, 0])\}.$$

Step 7: Compute the distance between each alternative and the ideal solutions (Table 6.28).
Step 8: Calculate the relative closeness to the ideal solutions (Table 6.28).
Step 9: Rank alternatives according to CC_i values, from highest to lowest (Table 6.28).
The ranking obtained is similar to the one obtained using the method of Chen and Lee (2010); the method proposed by Chen and Hong (2014) fails to distinguish A_2 and A_3.

It should be noted that normalization is with respect to the maximal element possible, which is 10, not the maximal element achieved; otherwise the ranking would invert A_2 and A_3, which is not expected given the individual assessments (Table 6.23).

CONCLUSION

This chapter reviewed TOPSIS in the fuzzy context. It presented precursors approaches with relevant examples. A MATLAB algorithm was proposed for the most widely recognized extension into the fuzzy context. The chapter also discussed frequently used distances, ideal solutions, normalization, and aggregation in fuzzy TOPSIS.

Even if the proposed method of reaching consensus requires fewer iterations than the original, it remains cumbersome. However, the results obtained better fit the definition of TOPSIS, because each alternative dominates the next in being both closer to the positive ideal solution and further from the negative ideal solution.

The chapter also discussed extensions of TOPSIS using type 1 and type 2 fuzzy sets with relevant adaptations.

REFERENCES

Ashtiani, Behzad, Farzad Haghighirad, Ahmad Makui, & Golam ali Montazer (2009). "Extension of Fuzzy TOPSIS Method Based on Interval-Valued Fuzzy Sets." *Applied Soft Computing*, 9(2), 457–461. https://doi.org/10.1016/j.asoc.2008.05.005.

Buyukozkan, Gulcin, Ismail Burak Parlak, & A. Cagri Tolga (2014). "Employing an Interval Type-2 Fuzzy Topsis Method for Knowledge Management Tool Evaluation." In *Decision Making and Soft Computing*, 9, 55–61. World Scientific Proceedings Series on Computer Engineering and Information Science, Volume 9. WORLD SCIENTIFIC. https://doi.org/10.1142/9789814619998_0012.

Büyüközkan, Gülçin, Ismail Burak Parlak, & A. Cagri Tolga (2016). "Evaluation of Knowledge Management Tools by Using An Interval Type-2 Fuzzy TOPSIS Method." *International Journal of Computational Intelligence Systems*, 9(5), 812–826. https://doi.org/10.1080/18756891.2016.1237182.

Cables, E., M. T. Lamata, & J. L. Verdegay (2016). "RIM-Reference Ideal Method in Multicriteria Decision Making." *Information Sciences*, 337–338(April), 1–10. https://doi.org/10.1016/j.ins.2015.12.011.

Cabrerizo, Francisco Javier, Ignacio Javier Pérez, Francisco Chiclana, & Enrique Herrera-Viedma (2017). "Group Decision Making: Consensus Approaches Based on Soft Consensus Measures." In *Fuzzy Sets, Rough Sets, Multisets and Clustering*, edited by Vicenç Torra, Anders Dahlbom, and Yasuo Narukawa, (pp. 307–321). Studies in Computational Intelligence. Cham: Springer International Publishing. https://doi.org/10.1007/978-3-319-47557-8_18.

Cengiz Toklu, Merve (2018). "Interval Type-2 Fuzzy TOPSIS Method for Calibration Supplier Selection Problem: A Case Study in an Automotive Company." *Arabian Journal of Geosciences*, 11(13), 341. https://doi.org/10.1007/s12517-018-3707-z.

Cha, Youngpil, & Mooyoung Jung (2003). "Satisfaction Assessment of Multi-Objective Schedules Using Neural Fuzzy Methodology." *International Journal of Production Research*, 41(8), 1831–1849. https://doi.org/10.1080/1352816031000074937.

Chen, Chen-Tung (2000). "Extensions of the TOPSIS for Group Decision-Making under Fuzzy Environment." *Fuzzy Sets and Systems*, *114*(1), 1–9. https://doi.org/10.1016/S0165-0114(97)00377-1.

Chen, Chen-Tung, Ching-Torng Lin, & Sue-Fn Huang (2006). "A Fuzzy Approach for Supplier Evaluation and Selection in Supply Chain Management." *International Journal of Production Economics*, *102*(2), 289–301.

Chen, Shan Huo, & Hsieh, Chih Hsun (1999). "Ranking Generalized Fuzzy Number with Graded Mean Integration Representation." Proceedings of the Eighth International Conference of Fuzzy Sets and Systems Association World Congress, vol. 2, 551–555.

Chen, Shan-Huo (1985). "Ranking Fuzzy Numbers with Maximizing Set and Minimizing Set." *Fuzzy Sets and Systems*, *17*(2), 113–129. https://doi.org/10.1016/0165-0114(85)90050-8.

Chen, Shu-Jen, & Ching-Lai Hwang (1992). "Fuzzy Multiple Attribute Decision Making Methods." In *Fuzzy Multiple Attribute Decision Making: Methods and Applications*, edited by Shu-Jen Chen, & Ching-Lai Hwang, (pp. 289–486). Lecture Notes in Economics and Mathematical Systems. Berlin, Heidelberg: Springer. https://doi.org/10.1007/978-3-642-46768-4_5.

Chen, Shyi-Ming (1996). "New Methods for Subjective Mental Workload Assessment and Fuzzy Risk Analysis." *Cybernetics and Systems*, *27*(5), 449–472. https://doi.org/10.1080/019697296126417.

Chen, Shyi-Ming, & Jia-An Hong (2014). "Fuzzy Multiple Attributes Group Decision-Making Based on Ranking Interval Type-2 Fuzzy Sets and the TOPSIS Method." *IEEE Transactions on Systems, Man, and Cybernetics: Systems*, *44*(12), 1665–1673. https://doi.org/10.1109/TSMC.2014.2314724.

Chen, Shyi-Ming, & Li-Wei Lee (2010). "Fuzzy Multiple Attributes Group Decision-Making Based on the Interval Type-2 TOPSIS Method." *Expert Systems with Applications*, *37*(4), 2790–2798. https://doi.org/10.1016/j.eswa.2009.09.012.

Chu, Ta-Chung (2002). "Facility Location Selection Using Fuzzy TOPSIS under Group Decisions." *International Journal of Uncertainty, Fuzziness and Knowledge-Based Systems*, *10*(06), 687–701. https://doi.org/10.1142/S0218488502001739.

Chu, T.-C., & Y.-C. Lin (2003). "A Fuzzy TOPSIS Method for Robot Selection." *The International Journal of Advanced Manufacturing Technology*, *21*(4), 284–290. https://doi.org/10.1007/s001700300033.

Cross, Valerie V., & Thomas A. Sudkamp (2002). *Similarity and Compatibility in Fuzzy Set Theory: Assessment and Applications*. 1st ed. Studies in Fuzziness and Soft Computing 93. Berlin, Heidelberg: Physica-Verlag Heidelberg. //www.springer.com/us/book/9783790814583.

Ding, Jie, Ze-shui Xu, & Hu-chang Liao (2017). "Consensus-Reaching Methods for Hesitant Fuzzy Multiple Criteria Group Decision Making with Hesitant Fuzzy Decision Making Matrices." *Frontiers of Information Technology & Electronic Engineering*, *18*(11), 1679–1692. https://doi.org/10.1631/FITEE.1601546.

Dymova, Ludmila, Pavel Sevastjanov, & Anna Tikhonenko (2015). "An Interval Type-2 Fuzzy Extension of the TOPSIS Method Using Alpha Cuts." *Knowledge-Based Systems*, *83*(July), 116–127. https://doi.org/10.1016/j.knosys.2015.03.014.

Dymova, Ludmila, Pavel Sevastjanov, & Anna Tikhonenko (2016). "The TOPSIS Method in the Interval Type-2 Fuzzy Setting." In *Parallel Processing and Applied Mathematics*, edited by Roman Wyrzykowski, Ewa Deelman, Jack Dongarra, Konrad Karczewski, Jacek Kitowski, & Kazimierz Wiatr, (pp. 445–454). Lecture Notes in Computer Science. Springer International Publishing.

El Alaoui, Mohamed (2018). "SMART Grid Evaluation Using Fuzzy Numbers and TOPSIS." *IOP Conference Series: Materials Science and Engineering*, *353*(1), 012019. https://doi.org/10.1088/1757-899X/353/1/012019.

El Alaoui, Mohamed (2020). "A Fuzzy Multiplicative Performance Indicator to Measure Circular Economy Efficiency" *International Journal of Mathematical, Engineering and Management Sciences* 5(6): 1118–1127.

El Alaoui, Mohamed, Hussain Ben-azza, & Khalid El Yassini (2018). "Optimal Weighting Method for Interval-Valued Intuitionistic Fuzzy Opinions." *Notes on Intuitionistic Fuzzy Sets*, 24(3), 106–110. https://doi.org/10.7546/nifs.2018.24.3.106-110.

El Alaoui, Mohamed, Hussain Ben-azza, & Khalid El Yassini (2019). "Achieving Consensus in Interval Valued Intuitionistic Fuzzy Environment." *Procedia Computer Science*, The Second International Conference On Intelligent Computing In Data Sciences, ICDS2018, 148(January), 218–225. https://doi.org/10.1016/j.procs.2019.01.064.

El Alaoui, Mohamed, & Khalid El Yassini (2020). "Fuzzy Similarity Relations in Decision Making." *Handbook of Research on Emerging Applications of Fuzzy Algebraic Structures*, 369–385. https://doi.org/10.4018/978-1-7998-0190-0.ch020.

El Alaoui, Mohamed, Khalid El Yassini, & Hussain Ben-azza (2019). "Type 2 Fuzzy TOPSIS for Agriculture MCDM Problems." *International Journal of Sustainable Agricultural Management and Informatics*, 5(2/3), 112–130. https://doi.org/10.1504/IJSAMI.2019.101672.

García, J. C. F. (2012). "A General Model for Linear Programming with Interval Type-2 Fuzzy Technological Coefficients." In 2012 Annual Meeting of the North American Fuzzy Information Processing Society (NAFIPS), 1–4. https://doi.org/10.1109/NAFIPS.2012.6291064.

Herrera-Viedma, Enrique, Francisco Javier Cabrerizo, Janusz Kacprzyk, & Witold Pedrycz (2014). "A Review of Soft Consensus Models in a Fuzzy Environment." *Information Fusion*, Special Issue: Information Fusion in Consensus and Decision Making, 17(May), 4–13. https://doi.org/10.1016/j.inffus.2013.04.002.

Hussian, Zahid, & Miin-Shen Yang (2019). "Distance and Similarity Measures of Pythagorean Fuzzy Sets Based on the Hausdorff Metric with Application to Fuzzy TOPSIS." *International Journal of Intelligent Systems* 34(10), 2633–2654. https://doi.org/10.1002/int.22169.

Ilieva, Galina (2016). "TOPSIS Modification with Interval Type-2 Fuzzy Numbers." *Cybernetics and Information Technologies*, 16(2), 60–68. https://doi.org/10.1515/cait-2016-0020.

Jahanshahloo, G. R., F. Hosseinzadeh Lotfi, & M. Izadikhah (2006). "An Algorithmic Method to Extend TOPSIS for Decision-Making Problems with Interval Data." *Applied Mathematics and Computation*, 175(2), 1375–1384. https://doi.org/10.1016/j.amc.2005.08.048.

Jin, Yaochu (2003). *Advanced Fuzzy Systems Design and Applications*. 1st Edition. Studies in Fuzziness and Soft Computing 112. Heidelberg: Physica-Verlag. https://doi.org/10.1007/978-3-7908-1771-3.

Kahraman, Cengiz, Ihsan Kaya, Sezi Çevik, Nüfer Yasin Ates, & Murat Gülbay (2008). "Fuzzy Multi-Criteria Evaluation of Industrial Robotic Systems Using Topsis." In *Fuzzy Multi-Criteria Decision Making: Theory and Applications with Recent Developments*, edited by Cengiz Kahraman, (pp. 159–186). Springer Optimization and Its Applications. Boston, MA: Springer US. https://doi.org/10.1007/978-0-387-76813-7_6.

Kaufmann, Arnold, & Madan M. Gupta (1988). *Fuzzy Mathematical Models in Engineering and Management Science*. North-Holland.

Kóczy T., László, & Domonkos Tikk (2000). *Fuzzy rendszerek*. Budapest: Typotex.

Lee, E. S., & R.-J. Li (1988). "Comparison of Fuzzy Numbers Based on the Probability Measure of Fuzzy Events." *Computers & Mathematics with Applications*, 15(10), 887–896. https://doi.org/10.1016/0898-1221(88)90124-1.

Lee, Hsuan-Shih (2002). "Optimal Consensus of Fuzzy Opinions under Group Decision Making Environment." *Fuzzy Sets and Systems*, 132(3), 303–315. https://doi.org/10.1016/S0165-0114(02)00056-8.

Li, Yan-Lai, Rui Wang, & Kwai-Sang Chin (2019). "New Failure Mode and Effect Analysis Approach Considering Consensus under Interval-Valued Intuitionistic Fuzzy Environment." *Soft Computing*, January. https://doi.org/10.1007/s00500-018-03706-5.

Liang, Gin-Shuh (1999). "Fuzzy MCDM Based on Ideal and Anti-Ideal Concepts." *European Journal of Operational Research*, *112*(3), 682–691. https://doi.org/10.1016/S0377-2217(97)00410-4.

Liao, T. Warren (2015). "Two Interval Type 2 Fuzzy TOPSIS Material Selection Methods." *Materials & Design*, *88*(December), 1088–1099. https://doi.org/10.1016/j.matdes.2015.09.113.

Liou, Tian-Shy, & Mao-Jiun J. Wang (1992). "Ranking Fuzzy Numbers with Integral Value." *Fuzzy Sets and Systems*, *50*(3), 247–255. https://doi.org/10.1016/0165-0114(92)90223-Q.

Liu, Juan, Felix T. S. Chan, Ya Li, Yajuan Zhang, & Yong Deng (2012). "A New Optimal Consensus Method with Minimum Cost in Fuzzy Group Decision." *Knowledge-Based Systems* *35*(November), 357–360. https://doi.org/10.1016/j.knosys.2012.04.015.

López-Ortega, Omar, & Félix Castro-Espinoza (2018). "Fuzzy Similarity Metrics and Their Application to Consensus Reaching in Group Decision Making." *Journal of Intelligent & Fuzzy Systems* Preprint (Preprint): 1–10. https://doi.org/10.3233/JIFS-18508.

Luukka, Pasi, & Mikael Collan (2016). "Histogram Ranking with Generalised Similarity-Based TOPSIS Applied to Patent Ranking." *International Journal of Operational Research*, *25*(4), 437–448. https://doi.org/10.1504/IJOR.2016.075290.

Madi, Elissa Nadia, Jonathan M. Garibaldi, & Christian Wagner (2015). "A Comparison between Two Types of Fuzzy TOPSIS Method." In 2015 IEEE International Conference on Systems, Man, and Cybernetics, 291–297. https://doi.org/10.1109/SMC.2015.63.

Madi, Elissa Nadia, Jonathan M. Garibaldi, & Christian Wagner (2016). "An Exploration of Issues and Limitations in Current Methods of TOPSIS and Fuzzy TOPSIS." In 2016 IEEE International Conference on Fuzzy Systems (FUZZ-IEEE), 2098–2105. Vancouver, BC, Canada: IEEE. https://doi.org/10.1109/FUZZ-IEEE.2016.7737950.

Madi, Elissa Nadia, Jonathan M. Garibaldi, & Christian Wagner (2017). "Exploring the Use of Type-2 Fuzzy Sets in Multi-Criteria Decision Making Based on TOPSIS." In 2017 IEEE International Conference on Fuzzy Systems (FUZZ-IEEE), 1–6. https://doi.org/10.1109/FUZZ-IEEE.2017.8015664.

Mahdavi, Iraj, Nezam Mahdavi-Amiri, Armaghan Heidarzade, & Rahele Nourifar (2008). "Designing a Model of Fuzzy TOPSIS in Multiple Criteria Decision Making." *Applied Mathematics and Computation*, Includes Special issue on Modeling, Simulation, and Applied Optimization (ICMSAO-07, *206*(2), 607–617. https://doi.org/10.1016/j.amc.2008.05.047.

Mei, Yanlan, & Kefan Xie (2019). "An Improved TOPSIS Method for Metro Station Evacuation Strategy Selection in Interval Type-2 Fuzzy Environment." *Cluster Computing*, *22*(2), 2781–2792. https://doi.org/10.1007/s10586-017-1499-7.

Mendel, Jerry M. (2017). *Uncertain Rule-Based Fuzzy Systems: Introduction and New Directions*. 2nd ed. Springer International Publishing. //www.springer.com/us/book/9783319513690.

Mokhtarian, M. N. (2015). "A Note on 'Extension of Fuzzy TOPSIS Method Based on Interval-Valued Fuzzy Sets.'" *Applied Soft Computing*, *26*(January), 513–514. https://doi.org/10.1016/j.asoc.2014.10.013.

Negi, Devendra Singh (1989). "Fuzzy Analysis and Optimization." Department of Industrial Engineering: Kansas State University.

Nehi, Hassan Mishmast, & Abazar Keikha (2016). "TOPSIS and Choquet Integral Hybrid Technique for Solving MAGDM Problems with Interval Type-2 Fuzzy Numbers." *Journal of Intelligent & Fuzzy Systems*, *30*(3), 1301–1310. https://doi.org/10.3233/IFS-152044.

Otheman, Adawiyah, & Lazim Abdullah (2014). "A New Concept of Similarity Measure for IT2FS TOPSIS and Its Use in Decision Making." *AIP Conference Proceedings*, *1602*(1), 608–614. https://doi.org/10.1063/1.4882547.

Peng, Xindong, & Jingguo Dai (2018). "Approaches to Single-Valued Neutrosophic MADM Based on MABAC, TOPSIS and New Similarity Measure with Score Function." *Neural Computing and Applications*, *29*(10), 939–954. https://doi.org/10.1007/s00521-016-2607-y.

Perez, Ignacio Javier, Francisco Javier Cabrerizo, Sergio Alonso, Francisco Chiclana, & Enrique Herrera-Viedma (2016). "Soft Consensus Models in Group Decision Making." In *Fuzzy Logic and Information Fusion: To Commemorate the 70th Birthday of Professor Gaspar Mayor*, edited by Tomasa Calvo Sánchez, & Joan Torrens Sastre, (pp. 135–153). Studies in Fuzziness and Soft Computing. Cham: Springer International Publishing. https://doi.org/10.1007/978-3-319-30421-2_10.

Ploskas, Nikolaos, & Jason Papathanasiou (2019). "A Decision Support System for Multiple Criteria Alternative Ranking Using TOPSIS and VIKOR in Fuzzy and Nonfuzzy Environments." *Fuzzy Sets and Systems*, Theme: Preference, Decision, Optimization, *377*(December), 1–30. https://doi.org/10.1016/j.fss.2019.01.012.

Rani, Pratibha, Arunodaya Raj Mishra, Ghasem Rezaei, Huchang Liao, & Abbas Mardani (2020). "Extended Pythagorean Fuzzy TOPSIS Method Based on Similarity Measure for Sustainable Recycling Partner Selection." *International Journal of Fuzzy Systems*, *22*(2), 735–747. https://doi.org/10.1007/s40815-019-00689-9.

Roghanian, E., J. Rahimi, & A. Ansari (2010). "Comparison of First Aggregation and Last Aggregation in Fuzzy Group TOPSIS." *Applied Mathematical Modelling*, *34*(12), 3754–3766. https://doi.org/10.1016/j.apm.2010.02.039.

Sadi-Nezhad, Soheil, & Kaveh Khalili Damghani (2010). "Application of a Fuzzy TOPSIS Method Base on Modified Preference Ratio and Fuzzy Distance Measurement in Assessment of Traffic Police Centers Performance." *Applied Soft Computing*, Optimisation Methods & Applications in Decision-Making Processes, *10*(4), 1028–1039. https://doi.org/10.1016/j.asoc.2009.08.036.

Sang, Xiuzhi, & Xinwang Liu (2016). "An Analytical Solution to the TOPSIS Model with Interval Type-2 Fuzzy Sets." *Soft Computing*, *20*(3), 1213–1230. https://doi.org/10.1007/s00500-014-1584-2.

Santini, Simone, & Ramesh Jain (1997). "Similarity Is a Geometer." *Multimedia Tools Appl*, *5*(3), 277–306. https://doi.org/10.1023/A:1009651725256.

Sharaf, Iman Mohamad (2018). "TOPSIS with Similarity Measure for MADM Applied to Network Selection." *Computational and Applied Mathematics*, *37*(4), 4104–4121. https://doi.org/10.1007/s40314-017-0556-4.

Sharaf, Iman Mohamad (2019). "An Interval Type-2 Fuzzy TOPSIS Using the Extended Vertex Method for MAGDM." *SN Applied Sciences*, *2*(1), 87. https://doi.org/10.1007/s42452-019-1825-1.

Shyi-Ming Chen, & Jia-An Hong (2014). "A New Method for Fuzzy Multiple Attributes Group Decision Making Based on Interval Type-2 Fuzzy Sets and the TOPSIS Method." In *2014 International Conference on Machine Learning and Cybernetics*, *1*, 338–344. https://doi.org/10.1109/ICMLC.2014.7009139.

Szmidt, Eulalia (2014). *Distances and Similarities in Intuitionistic Fuzzy Sets*. Studies in Fuzziness and Soft Computing. Springer International Publishing. //www.springer.com/la/book/9783319016399.

Tan, Y., Z. Cai, and H. Qi (2010). "A Process-Based Performance Analysis for Closed-Loop Agriculture Supply Chain." In *2010 International Conference on Intelligent System Design and Engineering Application*, *1*, 145–149. https://doi.org/10.1109/ISDEA.2010.91.

Triantaphyllou, Evangelos, & Chi-Tun Lin (1996). "Development and Evaluation of Five Fuzzy Multiattribute Decision-Making Methods." *International Journal of Approximate Reasoning*, *14*(4), 281–310. https://doi.org/10.1016/0888-613X(95)00119-2.

Tsaur, Sheng-Hshiung, Te-Yi Chang, & Chang-Hua Yen (2002). "The Evaluation of Airline Service Quality by Fuzzy MCDM." *Tourism Management*, *23*(2), 107–115. https://doi.org/10.1016/S0261-5177(01)00050-4.

Wang, Huidong, Jinli Yao, Jun Yan, & Mingguang Dong (2019). "An Extended TOPSIS Method Based on Gaussian Interval Type-2 Fuzzy Set." *International Journal of Fuzzy Systems*, *21*(6), 1831–1843. https://doi.org/10.1007/s40815-019-00670-6.

Wang, Ying-Ming, & Taha M. S. Elhag (2006). "Fuzzy TOPSIS Method Based on Alpha Level Sets with an Application to Bridge Risk Assessment." *Expert Systems with Applications*, *31*(2), 309–319. https://doi.org/10.1016/j.eswa.2005.09.040.

Wu, Jian, Francisco Chiclana, & Enrique Herrera-Viedma (2015). "Trust Based Consensus Model for Social Network in an Incomplete Linguistic Information Context." *Applied Soft Computing*, *35*(October), 827–839. https://doi.org/10.1016/j.asoc.2015.02.023.

Wu, Tong, Xinwang Liu, & Fang Liu (2018). "An Interval Type-2 Fuzzy TOPSIS Model for Large Scale Group Decision Making Problems with Social Network Information." *Information Sciences*, *432*(March), 392–410. https://doi.org/10.1016/j.ins.2017.12.006.

Xu, Yejun, Hao Sun, & Huimin Wang (2016). "Optimal Consensus Models for Group Decision Making under Linguistic Preference Relations." *International Transactions in Operational Research*, *23*(6), 1201–1228. https://doi.org/10.1111/itor.12154.

Xu, Yejun, & Huimin Wang (2013). "Optimal Weight Determination and Consensus Formation under Fuzzy Linguistic Environment." *Procedia Computer Science*, First International Conference on Information Technology and Quantitative Management, *17*(January), 482–489. https://doi.org/10.1016/j.procs.2013.05.062.

Yang, Taho, & Chih-Ching Hung (2007). "Multiple-Attribute Decision Making Methods for Plant Layout Design Problem." *Robotics and Computer-Integrated Manufacturing*, *23*(1), 126–137. https://doi.org/10.1016/j.rcim.2005.12.002.

Yang, Yu-Yao, Xin-Wang Liu, & Fang Liu (2020). "Trapezoidal Interval Type-2 Fuzzy TOPSIS Using Alpha-Cuts." *International Journal of Fuzzy Systems*, *22*(1), 293–309. https://doi.org/10.1007/s40815-019-00777-w.

Yue, Weichao, Weihua Gui, Xiaofang Chen, Zhaohui Zeng, & Yongfang Xie (2019). "Knowledge Representation and Reasoning Using Self-Learning Interval Type-2 Fuzzy Petri Nets and Extended TOPSIS." *International Journal of Machine Learning and Cybernetics*, *10*(12), 3499–3520. https://doi.org/10.1007/s13042-019-00940-7.

Yuen, Kevin Kam Fung (2014). "Combining Compound Linguistic Ordinal Scale and Cognitive Pairwise Comparison in the Rectified Fuzzy TOPSIS Method for Group Decision Making." *Fuzzy Optimization and Decision Making*, *13*(1), 105–130. https://doi.org/10.1007/s10700-013-9168-7.

Zamri, Nurnadiah, & Lazim Abdullah (2014). "Flood Control Project Selection Using an Interval Type-2 Entropy Weight with Interval Type-2 Fuzzy TOPSIS." *AIP Conference Proceedings*, *1602*(1), 62–68. https://doi.org/10.1063/1.4882467.

Zang, Tianlei, Lu Qiu, Zhengyou He, & Qingquan Qian (2016). "A Group Evaluation Method for Science Popularization Using Generalized Fuzzy Similarity and TOPSIS." In *2016 IEEE International Conference on Mechatronics and Automation*, 2487–2493. https://doi.org/10.1109/ICMA.2016.7558957.

Zhang, G., Y. Dong, Y. Xu, & H. Li (2011). "Minimum-Cost Consensus Models Under Aggregation Operators." *IEEE Transactions on Systems, Man, and Cybernetics - Part A: Systems and Humans*, *41*(6), 1253–1261. https://doi.org/10.1109/TSMCA.2011.2113336.

Zhang, Guangquan, & Jie Lu (2003). "An Integrated Group Decision-Making Method Dealing with Fuzzy Preferences for Alternatives and Individual Judgments for Selection Criteria." *Group Decision and Negotiation*, *12*(6), 501–515. https://doi.org/10.1023/B:GRUP.0000004197.04668.cf.

Zhang, Xiaolu, & Zeshui Xu (2015). "Soft Computing Based on Maximizing Consensus and Fuzzy TOPSIS Approach to Interval-Valued Intuitionistic Fuzzy Group Decision Making." *Applied Soft Computing*, *26*(January), 42–56. https://doi.org/10.1016/j.asoc.2014.08.073.

Zhao, Haoran, Huiru Zhao, & Sen Guo (2017). "Evaluating the Comprehensive Benefit of Eco-Industrial Parks by Employing Multi-Criteria Decision Making Approach for Circular Economy." *Journal of Cleaner Production*, *142*(January), 2262–2276. https://doi.org/10.1016/j.jclepro.2016.11.041.

Zhao, Renhong, & Rakesh Govind (1991). "Algebraic Characteristics of Extended Fuzzy Numbers." *Information Sciences*, *54*(1), 103–130. https://doi.org/10.1016/0020-0255(91)90047-X.

7 Intuitionistic Fuzzy TOPSIS

INTRODUCTION

Fuzzy TOPSIS extensions with type 1 fuzzy sets, detailed in the previous chapter, are the largest group of extensions of fuzzy TOPSIS; but the second largest is intuitionistic fuzzy TOPSIS extensions, which are discussed in this chapter (Salih et al., 2019).

A Scopus database search on April 10, 2020, for the terms ("intuitionistic" AND "fuzzy" AND "TOPSIS") OR ("intuitionistic" AND "fuzziness" AND "TOPSIS") in the period 2006–2019 produced 410 results, as shown in Figure 7.1.

Before getting to intuitionistic TOPSIS extensions, though, there are two main problems to deal with in the intuitionistic fuzzy context: aggregation operators and the distances to use.

INFORMATION INTEGRATION

"What is life? The usufruct of an aggregation of molecules." Edmond and Jules de Goncourt

The development of an aggregation operator is doubly useful. In the first place, the amount of data available is astronomical; in a Google search for "intuitionistic fuzzy TOPSIS" in March 2020, the number of responses exceeded 80,000. In the second place, as the assessment step involves several decision makers with different backgrounds and perspectives, the search for consensus is pressing. (A more elaborate discussion of consensus was given in Chapter 6.)

Before going any further, it would be appropriate to position the problem in a more global perspective. Torra and Narukawa (2007) distinguish data integration from data fusion and aggregation operators. While data integration is the use of data from multiple sources (or the same source but obtained at different times) to accomplish a particular task, data fusion is the real process of combining various data into a single representative one. Therefore, fusion involves particular mathematical functions, algorithms, methods, and procedures for combining data. Consequently, data fusion is one of the processes embedded in a data-integration architecture (see Figure 7.2 for an overall diagram):

- Acquisition: the collection of information from available sources. This can also be called detection. In order to obtain good-quality information, a data model is required to measure the quality of information.
- Preprocessing: the preparation of data for the fusion stage, making it computable and commensurable. It can take many forms, ranging from noise

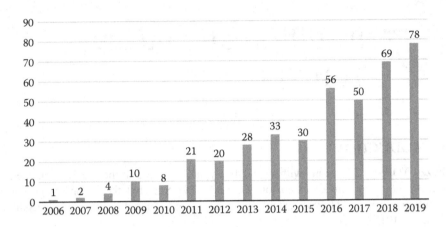

FIGURE 7.1 Intuitionistic Fuzzy TOPSIS in the Scopus Database, 2006–2019.

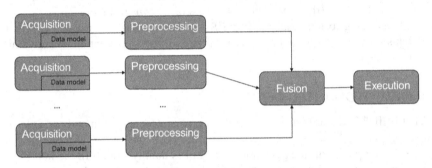

FIGURE 7.2 General Architecture of Data Fusion (Torra & Narukawa, 2007).

reduction to edge detection. In order to avoid a vague definition leading to confusion between preprocessing and fusion, we can state the following: as long as the data used is from a single source (as shown in Figure 7.2), we are talking about pretreatment. In what follows, we simply assume that the preprocessing step standardizes the data collected into the unity interval [0, 1].
- Fusion: contrary to preprocessing, acts on all data. It merges all the data into a small number, typically only one, of the same nature as the entries. For example, the result of merging several opinions is also an opinion. However, some systems may have a different approach.
- Execution: selection of the appropriate procedure based on the outcome of the fusion stage. There are two forms of procedures: application of action and interpretation of data.

In the context of multicriteria decision making, acquisition corresponds to the collection of alternative assessments according to each criterion and criterion weights by each decision maker using a linguistic variable. Pretreatment converts the linguistic variables into the unity interval [0, 1]. Then aggregation is carried out on the basis of a suitable operator to yield collective assessments, a second aggregation occurs to combine alternative performances according to each criterion, and finally a decision is made by selecting the alternative that is the best rated.

AGGREGATION FUNCTIONS

Aggregation consists of combining several numerical values $x = (x_1, \ldots x_n)$ into a single representative value by an aggregation function $f(x)$ (Grabisch et al., 2009). The aggregation function can be represented in several forms (Beliakov, Pradera, & Calvo, 2007):

- An algebraic function $f(x) = x_1 + 3x_2 + x_3$
- A graphical function (two-dimensional, three-dimensional, or contour)
- Verbal, in a sequence of steps (algorithm)
- A consultation table
- A solution to certain equations (algebraic, differential, or functional)
- A computer subprogram that returns a value y for any specified x

Traditionally, an aggregation function on a space of size n is defined by

$$f: [0, 1]^n \to [0, 1],$$

satisfying the boundary conditions (Gleb, Humberto, & Tomasa, 2016) $f(0, \ldots, 0) = 0$ and $f(1, \ldots, 1) = 1$ and monotonicity: $f(x_1, \ldots, x_n) \leq f(y_1, \ldots, y_n)$ if $x_1 \leq y_1; \ldots ; x_n \leq y_n$.

Aggregation operators could be divided into the following four main categories (Beliakov, Pradera, & Calvo, 2007).

COMPROMISE OPERATORS

An aggregation function f is called a compromise or an averaging function if

$$\min(x_1, \ldots, x_n) \leq f(x_1, \ldots, x_n) \leq \max(x_1, \ldots, x_n).$$

The best-known such functions are simple and weighted additive means (Chapter 4).

CONJUNCTIVE OPERATORS

An operator is called conjunctive if

$$f(x_1, \ldots, x_n) \leq \min(x_1, \ldots, x_n).$$

This matches the logical "AND": the end result will be high only if all the input arguments are.

DISJUNCTIVE OPERATORS

An operator is called disjunctive if

$$\max(x_1, \ldots, x_n) \leq f(x_1, \ldots, x_n).$$

This matches the logical "OR": it only takes one high input argument to yield a high end result.

MIXED OPERATORS

An aggregation function is said to be mixed if it does not belong to any of the other three classifications.

INTUITIONISTIC FUZZY AGGREGATION OPERATORS

Intuitionistic fuzzy aggregation operators are mainly derived from two specific sources (Xu & Cai, 2012): Intuitionistic Fuzzy Weighted Averaging (IFWA; Xu, 2007) and Intuitionistic Fuzzy Weighted Geometric (IFWG; Xu & Yager, 2006).

Let $\alpha_i = (\mu_i, \vartheta_i)$, with $1 \leq i \leq n$, be a collection of intuitionistic fuzzy numbers and ω_i be the relative weight of each α_i. The IFWA operator is defined from $[0, 1]^{2n}$ to $[0, 1]^2$ as

$$IFWA_\omega(\alpha_1, \alpha_2, \ldots, \alpha_n) = \left(1 - \prod_{i=1}^{n}(1-\mu_i)^{\omega_i}, \prod_{i=1}^{n} \vartheta_i^{\omega_i}\right). \quad (7.1)$$

Similarly, the IFWG is defined from $[0, 1]^{2n}$ to $[0, 1]^2$ as

$$IFWG_\omega(\alpha_1, \alpha_2, \ldots, \alpha_n) = \left(\prod_{i=1}^{n} \mu_i^{\omega_i}, 1 - \prod_{i=1}^{n}(1-\vartheta_i)^{\omega_i}\right). \quad (7.2)$$

The majority of the other intuitionist fuzzy aggregation operators rely on these two operators. However, noting that

$$IFWA_\omega(\alpha_1, \alpha_2, \ldots, \alpha_n) = \alpha_{IFWA}(\mu_{\alpha_{IFWA}}, \vartheta_{\alpha_{IFWA}})$$
$$\text{and } IFWG_\omega(\alpha_1, \alpha_2, \ldots, \alpha_n) = \alpha_{IFWG}(\mu_{\alpha_{IFWG}}, \vartheta_{\alpha_{IFWG}}),$$

if there exists an α_i such that $\mu_i = 1$, then $\mu_{\alpha_{IFWA}} = 1$ regardless of the other elements. And if there exists an α_i such that $\vartheta_i = 0$, then $\vartheta_{\alpha_{IFWA}} = 0$.

Similarly, the existence of any α_i such that $\mu_i = 0$ implies $\mu_{\alpha_{IFWG}} = 0$; and the existence of any α_i such that $\vartheta_i = 1$ implies $\vartheta_{\alpha_{IFWG}} = 1$.

To avoid this veto behavior, a Generalized IFWA (GIFWA) was proposed by El Alaoui and Ben-azza (2017) that can be adapted as follows:

$$GIFWA_\omega(\alpha_1, \alpha_2, \ldots, \alpha_n) = \left(1 - \frac{A_A - B}{C - B}, \frac{D_A - E}{F - E}\right), \tag{7.3}$$

where

$A_A = \prod_{i=1}^{n} (p_i + 1 - \mu_i)^{(1/n)}$, $B = \prod_{i=1}^{n} p_i^{(1/n)}$, $C = \prod_{i=1}^{n} (p_i + 1)^{(1/n)}$, $D_A = \prod_{i=1}^{n} (p_i + \vartheta_i)^{(1/n)}$, $E = \prod_{i=1}^{n} p_i^{(1/n)}$, and $F = \prod_{i=1}^{n} (p_i + 1)^{(1/n)}$, with p_i being inversely proportional to weights. Generally, $p_i = 1 - w_i$.

Similarly, the extension of IFWG into Generalized IFWG (GIFWG) is

$$GIFWG_\omega(\alpha_1, \alpha_2, \ldots, \alpha_n) = \left(\frac{A_G - B}{C - B}, 1 - \frac{D_G - E}{F - E}\right), \tag{7.4}$$

where

$A_G = \prod_{i=1}^{n} (p_i + \mu_i)^{(1/n)}$ and $D_G = \prod_{i=1}^{n} (p_i + 1 - \vartheta_i)^{(1/n)}$.

Generally, $p_i \in]0, 1]$; by fixing $p_i = 0$ for all i in Equations 7.3 (GIFWA) and 7.4 (GIFWG), we get back to, respectively, the IFWA (Equation 7.1) and IFWG (Equation 7.2) operators.

In addition to avoiding being absorbed by extreme values, the proposed operators are more acceptable from a practical point of view, especially in decision making, since the final assessment obtained would be that of a compromise operator. Development and generalizations in interval intuitionist fuzzy logic have been proposed by El Alaoui and Ben-azza (2017).

Let $\beta_i = ([a_i, b_i], [c_i, d_i])$ be a collection of interval valued intuitionistic fuzzy numbers. IFWA and IFWG have been respectively extended to the interval-valued intuitionistic fuzzy context by Interval-Valued IFWA (IVIFWA) and Interval-Valued IFWG (IVIFWG):

$$IVIFWA_\omega(\beta_1, \beta_2, \ldots, \beta_n) = \left(\left[1 - \prod_{i=1}^{n}(1 - a_i)^{\omega_i}, 1 - \prod_{i=1}^{n}(1 - b_i)^{\omega_i}\right], \left[\prod_{i=1}^{n} c_i^{\omega_i}, \prod_{i=1}^{n} d_i^{\omega_i}\right]\right) \tag{7.5}$$

$$IVIFWG_\omega(\beta_1, \beta_2, \ldots, \beta_n) = \left(\left[\prod_{i=1}^{n} a_i^{\omega_i}, \prod_{i=1}^{n} b_i^{\omega_i}\right], \left[1 - \prod_{i=1}^{n}(1 - c_i)^{\omega_i}, 1 - \prod_{i=1}^{n}(1 - d_i)^{\omega_i}\right]\right). \tag{7.6}$$

GIFWA has also been extended to Interval-Valued GIFWA (IVGIFWA; El Alaoui & Ben-azza, 2017):

$$IVGIFWA_\omega(\beta_1, \beta_2, \ldots, \beta_n) = ([aa, ba], [ca, da]) \tag{7.7}$$

where

$$aa = 1 - \frac{\prod_{i=1}^{n}(p_i + 1 - a_i)^{1/n} - \prod_{i=1}^{n} p_i^{1/n}}{\prod_{i=1}^{n}(p_i + 1)^{1/n} - \prod_{i=1}^{n} p_i^{1/n}}$$

$$ba = 1 - \frac{\prod_{i=1}^{n}(p_i + 1 - b_i)^{1/n} - \prod_{i=1}^{n} p_i^{1/n}}{\prod_{i=1}^{n}(p_i + 1)^{1/n} - \prod_{i=1}^{n} p_i^{1/n}}$$

$$ca = \frac{\prod_{i=1}^{n}(p_i + c_i)^{1/n} - \prod_{i=1}^{n} p_i^{1/n}}{\prod_{i=1}^{n}(p_i + 1)^{1/n} - \prod_{i=1}^{n} p_i^{1/n}}$$

$$da = \frac{\prod_{i=1}^{n}(p_i + d_i)^{1/n} - \prod_{i=1}^{n} p_i^{1/n}}{\prod_{i=1}^{n}(p_i + 1)^{1/n} - \prod_{i=1}^{n} p_i^{1/n}}$$

And GIFWG similarly has been extended to Interval-Valued GIFWG (IVGIFWG) as follows:

$$IVGIFWG_\omega(\beta_1, \beta_2, \ldots, \beta_n) = ([ag, bg], [cg, dg]), \tag{7.8}$$

where

$$ag = \frac{\prod_{i=1}^{n}(p_i + a_i)^{1/n} - \prod_{i=1}^{n} p_i^{1/n}}{\prod_{i=1}^{n}(p_i + 1)^{1/n} - \prod_{i=1}^{n} p_i^{1/n}}$$

$$bg = \frac{\prod_{i=1}^{n}(p_i + b_i)^{1/n} - \prod_{i=1}^{n} p_i^{1/n}}{\prod_{i=1}^{n}(p_i + 1)^{1/n} - \prod_{i=1}^{n} p_i^{1/n}}$$

$$cg = 1 - \frac{\prod_{i=1}^{n}(p_i + 1 - c_i)^{1/n} - \prod_{i=1}^{n} p_i^{1/n}}{\prod_{i=1}^{n}(p_i + 1)^{1/n} - \prod_{i=1}^{n} p_i^{1/n}}$$

$$dg = 1 - \frac{\prod_{i=1}^{n}(p_i + 1 - d_i)^{1/n} - \prod_{i=1}^{n} p_i^{1/n}}{\prod_{i=1}^{n}(p_i + 1)^{1/n} - \prod_{i=1}^{n} p_i^{1/n}}$$

Further discussions on intuitionistic aggregation operators are given by Xu (2012), Xu and Cai (2012), and Zhao et al. (2010).

INTUITIONISTIC FUZZY DISTANCES

Generally, in a classical set X the distance between two elements x_1 and x_2 fulfills the following conditions (Deza & Deza, 2006):

- $\forall \ x_1, x_2 \in X^2: \ d(x_1, x_2) \geq 0$ (positivity)
- $\forall \ x_1, x_2 \in X^2: \ d(x_1, x_2) = d(x_2, x_1)$ (reflexivity)
- $\forall \ x_1, x_2 \in X^2: \ d(x_1, x_2) = 0 \Leftrightarrow x_2 = x_1$ (separability)
- $\forall \ x_1, x_2 \in X^3: \ d(x_1, x_3) \leq d(x_1, x_2) + d(x_2, x_3)$ (triangular inequality).

In addition, in the fuzzy context distances are required to lie in the unity interval. Thus, with two intuitionistic fuzzy numbers α_1 and α_2 (Xu & Chen, 2008), $0 \leq d(\alpha_1, \alpha_2) \leq 1$.

According to Atanassov (1999), with intuitionistic fuzzy sets, two main distances can be used. Let F be a fuzzy set and $\alpha_1(\mu_1, \vartheta_1)$ and $\alpha_2(\mu_2, \vartheta_2)$ be two intuitionistic fuzzy numbers. The Hamming distance is defined as

$$d_H = \frac{1}{2}(|\mu_1 - \mu_2| + |\vartheta_1 - \vartheta_2|), \quad (7.9)$$

whereas the Euclidean intuitionistic fuzzy distance is

$$d_E = \sqrt{\frac{1}{2}((\mu_1 - \mu_2)^2 + (\vartheta_1 - \vartheta_2)^2)}. \quad (7.10)$$

Other authors also include hesitancy (Szmidt, 2014), and thus the Hamming distance becomes

$$d_H = \frac{1}{2}(|\mu_1 - \mu_2| + |\vartheta_1 - \vartheta_2| + |\pi_1 - \pi_2|), \quad (7.11)$$

and the Euclidean distance becomes

$$d_E = \sqrt{\frac{1}{2}((\mu_1 - \mu_2)^2 + (\vartheta_1 - \vartheta_2)^2 + (\pi_1 - \pi_2)^2)}, \quad (7.12)$$

For interval-valued intuitionistic fuzzy values $\beta_i = ([a_i, b_i], [c_i, d_i])$, the Hamming distance (Equation 7.9) becomes (Atanassov, 1999)

$$d_H(\beta_1, \beta_2) = \frac{1}{4}(|a_1 - a_2| + |b_1 - b_2| + |c_1 - c_2| + |d_1 - d_2|), \quad (7.13)$$

while the Euclidean distance (Equation 7.10) becomes

$$d_E = \sqrt{\frac{1}{4}((a_1 - a_2)^2 + (b_1 - b_2)^2 + (c_1 - c_2)^2 + (d_1 - d_2)^2)}, \quad (7.14)$$

INTUITIONISTIC FUZZY CONSENSUS

Several consensus-reaching approaches have been proposed in the intuitionistic fuzzy context (Jiang, Xu, & Yu, 2013; Liao et al., 2016; Szmidt & Kacprzyk, 2003; Wu & Chiclana, 2014; Xu et al., 2016). While some are preference oriented (Chu et al., 2016; Wan, Wang, & Dong, 2018; Zhang et al., 2020; Zhang & Pedrycz, 2018, 2019), and thus adapted for outranking methods, others are more fit for value-measurement approaches (Li, Wang, & Chin, 2019; Wu, Liu, & Liang 2015; Zhang, Li, & Xu, 2014), including TOPSIS (Zhang & Xu, 2015).

El Alaoui, Ben-azza, and El Yassini (2018) proposed a first attempt at adapting Lee's algorithm (Chapter 6) to the interval-valued intuitionistic fuzzy context. They used a modified Hamming distance (Equation 7.13) combined with the IVIFWA operator (Equation 7.5), and deduced the similarity based on Santini's formula (Santini & Jain, 1997) of $S = 1 - D$.

El Alaoui, Ben-azza, and El yassini (2019) expanded previous work investigating combinations of the IVIFWA (Equation 7.5) and IVIFWG (Equation 7.6) aggregation operators and the two most-used interval-valued intuitionistic fuzzy metrics, the Hamming (Equation 7.13) and Euclidean (Equation 7.14) distances, in addition to different starting points. They concluded that the starting points have practically no influence on the final result, but starting near the consensus may reduce the number of iterations needed. They concluded that the combination of the Hamming distance and IVIFWA produces better results.

Modifying the similarity to become

$$S = 1 - \frac{\sqrt{((a_1 - a_2)^2 + (b_1 - b_2)^2 + (c_1 - c_2)^2 + (d_1 - d_2)^2)}}{4} \quad (7.15)$$

and using IVGIFWA (Equation 7.7) and IVGIFWG (Equation 7.8), Algorithm 6.3 becomes the following:

Algorithm 7.1

Step 1: Each decision maker k ($1 \leq k \leq K$) expresses their assessment using an interval-valued intuitionistic fuzzy number.

Step 2: Fix the initial aggregation weights satisfying $0 \leq g_k^{(0)} \leq 1$ and $\sum_{k=1}^{K} g_k^{(0)} = 1$. The iterations will be labeled $l = 0, 1, 2, \ldots$.

Step 3: Compute the aggregated consensus at iteration l by l IVGIFWA or IVGIFWG:

$$\beta^l = IVGIFWA_w(\beta_1, \ldots, \beta_K)$$

or

$$\beta^l = IVGIFWG_w(\beta_1, \ldots, \beta_K)$$

Step 4: Using the weights at iteration l $G^{(l)}(g_1^{(l)}, \ldots, g_K^{(l)})$, compute $G^{(l+1)}$ as follows:

$$g_k^{(l+1)} = \frac{(1/(c - S(\beta^l, \beta_k)))^{1/(t-1)}}{\sum_{k2}^{K} (1/(c - S(\beta^l, \beta_{k2})))^{1/(t-1)}}.$$

Step 5: If $\|G^{(l+1)} - G^{(l)}\| > \varepsilon = 10^{-6}$, set $l = l + 1$ and go to step 3; otherwise, stop.

To compare the effect of the IVGIFWA and IVGIFWG aggregation operators and similarity (Equation 7.15) on the final consensus, consider the example from El Alaoui, Ben-azza, and El Yassini (2019), in which three decision makers evaluate an alternative with $\beta_1 = ([0.22, 0.31], [0.23, 0.54])$, $\beta_2 = ([0.04, 0.21], [0.35, 0.46])$, and $\beta_3 = ([0.25, 0.27], [0.23, 0.4])$. After fixing $c = 1.5$, $t = 2$, and $\varepsilon = 10^{-6}$ (El Alaoui, Ben-azza, & El Yassini, 2018, 2019), the algorithm converges in only five iterations, whereas it requires six in the best cases for El Alaoui et al. (2019) and seven for El Alaoui et al. (2018). Table 7.1 details the results of the six combinations examined.

Tables 7.2–7.7 show the results of each combination.

The different starting points have no effect on the final results, since the first three combinations converge to the same result alongside the last three combinations. IVGIFWA results in a slightly better optimized function.

SOME INTUITIONISTIC FUZZY TOPSIS EXTENSIONS

Several extensions have been proposed in the literature for the TOPSIS algorithm in the intuitionistic (Shen et al., 2018; Yan & Jia, 2011; Ye, 2015; Yuan & He, 2012) and the interval-valued intuitionistic fuzzy contexts (De, Das, & Kar, 2019; Gupta et al., 2018; Park et al., 2011; Wang & Chen, 2017; Zhang & Yu, 2012; Ye, 2010).

Perhaps the first extension of TOPSIS to intuitionistic and interval-valued intuitionistic fuzzy contexts was proposed by Tan and Zhang (2006), who used a modified Euclidean distance and faithfully followed the steps of TOPSIS while introducing the needed adaptation to the intuitionistic context.

Yuan and He (2012) proposed a new distance and relative similarity to avoid unreliable results that may be produced by some classically used metrics. The closeness coefficients were based on the computed similarity.

Yan and Jia (2011) proposed a method based on linguistic variables in which attribute weights were determined through maximizing deviations, whereas decision-maker weights were crisp known numbers. The distance used was not normalized in the unity interval.

TABLE 7.1
Combinations Examined

Combination N⁰	Starting Point	Aggregation Operator	Iterations until Convergence	Optimized Function Value
1	(1, 0, 0)	IVGIFWA	5	0.168063694
2	(0, 1, 0)	IVGIFWA	5	0.168063694
3	(0, 0, 1)	IVGIFWA	5	0.168063694
4	(1, 0, 0)	IVGIFWG	5	0.168066239
5	(0, 1, 0)	IVGIFWG	5	0.168066239
6	(0, 0, 1)	IVGIFWG	5	0.168066239

TABLE 7.2
Combination 1

Iteration l	$g_1^{(l)}$	$g_2^{(l)}$	$g_3^{(l)}$	a^l	b^l	c^l	d^l
0	1	0	0				
1	0.334170525	0.339345906	0.326483569	0.12546692	0.18861045	0.45705547	0.65103582
2	0.334214919	0.331630188	0.334154893	0.1725207	0.26385588	0.26858709	0.46537178
3	0.334223011	0.331605199	0.33417179	0.17287167	0.2639646	0.26826782	0.46522654
4	0.334223045	0.331605119	0.334171836	0.17287274	0.26396502	0.2682668	0.46522641
5	0.334223045	0.331605119	0.334171836	0.17287274	0.26396502	0.2682668	0.46522641

TABLE 7.3
Combination 2

Iteration l	$g_1^{(l)}$	$g_2^{(l)}$	$g_3^{(l)}$	a^l	b^l	c^l	d^l
0	0	1	0	0.09240987	0.167278133	0.509650857	0.628225459
1	0.332076621	0.342446465	0.325476914	0.172393705	0.263790753	0.268717389	0.465346015
2	0.334209491	0.331640244	0.334150265	0.172871252	0.263964411	0.26826823	0.465226507
3	0.334222995	0.331605231	0.334171774	0.172872738	0.263965022	0.268266802	0.465226407
4	0.334223045	0.331605119	0.334171836	0.172872743	0.263965024	0.268266797	0.465226407
5	0.334223045	0.331605119	0.334171836	0.172872743	0.263965024	0.268266797	0.465226407

TABLE 7.4
Combination 3

Iteration l	$g_1^{(l)}$	$g_2^{(l)}$	$g_3^{(l)}$	a^l	b^l	c^l	d^l
0	0	0	1	0.13192127	0.179646816	0.457055467	0.608075995
1	0.333147154	0.339373835	0.327479011	0.172526911	0.263845635	0.268588055	0.465330105
2	0.334213835	0.331630219	0.334155946	0.172871674	0.26396459	0.268267821	0.465226498
3	0.33422301	0.331605199	0.334171791	0.17287274	0.263965023	0.268266801	0.465226407
4	0.334223045	0.331605119	0.334171836	0.172872743	0.263965024	0.268266797	0.465226407
5	0.334223045	0.331605119	0.334171836	0.172872743	0.263965024	0.268266797	0.465226407

TABLE 7.5
Combination 4

Iteration l	$g_1^{(l)}$	$g_2^{(l)}$	$g_3^{(l)}$	a^l	b^l	c^l	d^l
0	1	0	0	0.415051525	0.491998295	0.174666801	0.343180283
1	0.33926715	0.316554157	0.344178693	0.166019904	0.262886333	0.270667214	0.468000647
2	0.334129866	0.331974246	0.333895888	0.164691776	0.262459633	0.27112829	0.468055987
3	0.334096718	0.332050615	0.333852666	0.16468526	0.26245745	0.271130548	0.468055895
4	0.334096545	0.332050992	0.333852462	0.164685228	0.262457439	0.271130559	0.468055894
5	0.334096544	0.332050994	0.333852461	0.164685228	0.262457439	0.271130559	0.468055894

TABLE 7.6
Combination 5

Iteration l	$g_1^{(l)}$	$g_2^{(l)}$	$g_3^{(l)}$	a^l	b^l	c^l	d^l
0	0	1	0	0.24798378	0.44368843	0.20151524	0.31933168
1	0.33538834	0.32417459	0.34043707	0.16537885	0.26265582	0.27089608	0.46797796
2	0.3341122	0.33201181	0.33387598	0.16468858	0.26245854	0.2711294	0.46805589
3	0.33409663	0.3320508	0.33385257	0.16468524	0.26245744	0.27113055	0.46805589
4	0.33409654	0.33205099	0.33385246	0.16468523	0.26245744	0.27113056	0.46805589
5	0.33409654	0.33205099	0.33385246	0.16468523	0.26245744	0.27113056	0.46805589

Intuitionistic Fuzzy TOPSIS

TABLE 7.7
Combination 6

Iteration l	$g_1^{(l)}$	$g_2^{(l)}$	$g_3^{(l)}$	a^l	b^l	c^l	d^l
0	0	0	1	0.4296265	0.47473744	0.1746668	0.30392675
1	0.33827346	0.31641693	0.34530962	0.16604325	0.2628764	0.2706627	0.46795834
2	0.33412913	0.3319734	0.33389748	0.16469186	0.26245964	0.27112826	0.46805594
3	0.33409672	0.33205061	0.33385267	0.16468526	0.26245745	0.27113055	0.4680559
4	0.33409655	0.33205099	0.33385246	0.16468523	0.26245744	0.27113056	0.46805589
5	0.33409654	0.33205099	0.33385246	0.16468523	0.26245744	0.27113056	0.46805589

Another approach based on maximizing deviations was proposed by Shen et al. (2018), who discussed the situation where the weights are completely known and there is not a relative resolution model for each situation. The composite distance used could have negative values.

In the interval-valued intuitionistic fuzzy environment, various methods have tried to derive alternative weights using an optimization approach.

Wang and Chen (2017) used a similarity based on the weighted Hamming distance and the Hausdorff metric. This approach used linear programming where weights and attribute assessments are represented by interval-valued intuitionistic fuzzy values.

An extension proposed by Zhang and Yu (2012) aimed to derive weights through an optimization model with incomplete weight information. The mathematical programming approach proposed was based on cross entropy.

Park et al. (2011) considered attribute weights to be partially known, and used interval-valued intuitionistic fuzzy hybrid geometric aggregation operators to deduce the collective evaluation and a score function to defuzzify the fuzzy value obtained.

The extension proposed by De, Das, and Kar (2019) also supposed partially known weight information. The attribute assessments were expressed by probabilistic interval-valued intuitionistic hesitant fuzzy elements, while optimal weights were derived by linear programming.

Although the approach proposed by Ye (2010) presumed an interval-valued intuitionistic fuzzy context, criterion and decision-maker weights were supposed to be crisp. To the contrary, all the values used by Gupta et al. (2018) were expressed as interval-valued intuitionistic fuzzy numbers. They also discussed the differences between the parameters used in several methods, which can be reviewed in Table 7.8.

Li (2010) indicates that the membership and the nonmembership degrees can take any values in the membership and the nonmembership intervals, respectively.

RESOLUTION ALGORITHM

The Approach

Algorithm 7.2: As in previous chapters, let K decision makers ($1 \leq k \leq K$) assess m alternatives ($1 \leq i \leq m$) according to n criteria ($1 \leq j \leq n$).

Step 1: Each decision maker assesses each alternative using a linguistic variable $\tilde{\beta}_{ijk}$ (Table 7.9).

Step 2: Each decision maker assesses each criterion's importance using a linguistic variable $\tilde{w}_{jk}$ (Table 7.9).

Step 3: Using the individual alternative assessments $\tilde{\beta}_{ijk}([a_{\beta_{ijk}}, b_{\beta_{ijk}}], [c_{\beta_{ijk}}, d_{\beta_{ijk}}])$ and criterion weights $\tilde{w}_{jk}([a_{w_{jk}}, b_{w_{jk}}], [c_{w_{jk}}, d_{w_{jk}}])$, compute the collective alternative assessments $\tilde{\beta}_{ij}$ and criterion weights $\tilde{w}_j$ using Algorithm 7.1.

Since all interval-valued intuitionistic fuzzy values used are in the unity interval, no normalization is required.

TABLE 7.8
Parameters Used in Some TOPSIS Extensions in the Interval-Valued Intuitionistic Fuzzy Context (Gupta et al., 2018)

Reference	Decision-Making Process	Attribute Weights	Decision-Maker Weights	Mathematical Metric	Hesitation Parameter	Methodology
Wang and Chen (2017)	Single	IVIFN	Not applicable	Similarity degree comprising weighted Hamming distance and Hausdorff metric	Excluded	Extended TOPSIS
Ye (2010)	Group	Completely known and numeric	Completely known and numeric	Weighted Euclidean distance	Included	Extended TOPSIS
Wan and Dong (2014)	Group	IVIFN	Completely unknown	Similarity degree based on Hamming distance	Excluded	Possibility degree
Park et al. (2009)	Group	Partially/completely known	Completely known and numeric	Score and accuracy function	Excluded	Correlation coefficient
Wei (2010)	Group	Completely known and numeric	Completely known and numeric	Score and accuracy function	Excluded	Induced geometric aggregation operators
Wang and Liu (2013)	Group	Completely unknown	Not considered	Normalized and weighted Euclidean distance	Included	Extended LINMAP
Ye (2013)	Group	Completely unknown	Completely unknown	Weighted correlation coefficient	Excluded	Correlation coefficient
Zhang and Xu (2016)	Group	Completely unknown	Completely unknown	Utility values	Included	Utility theory

(Continued)

TABLE 7.8 (Continued)

Reference	Decision-Making Process	Attribute Weights	Decision-Maker Weights	Mathematical Metric	Hesitation Parameter	Methodology
Solairaju, Robinson, and Rethinakumar (2014)	Group	Completely unknown	Completely known and numeric	IIFOWA and IIFHA operators	Excluded	Entropy weights
Meng and Tang (2013)	Group	Not exactly known	Completely known and numeric	AIVIFCA and 2AGSAIVIFCA operators	Included	Cross entropy measure and Choquet integral
Gupta et al. (2018)	Group	IVIFN	IVIFN	Similarity degree comprising weighted Hamming distance and Hausdorff metric	Included	Integrated extended TOPSIS

TABLE 7.9
Linguistic Variables for Assessing Alternatives and Criterion Importance

Linguistic Variable for Alternatives	Linguistic Variable for Criteria	Interval-Valued Intuitionistic Fuzzy Number
Extremely low (EL)	Extremely poor (EP)	([0, 0], [1, 1])
Very low (VL)	Very poor (VP)	([0, 0.1], [0.8, 0.9])
Low (L)	Poor (P)	([0.1, 0.2], [0.7, 0.8])
Medium low (ML)	Medium poor (MP)	([0.2, 0.3], [0.6, 0.7])
Medium (M)	Fair (F)	([0.5, 0.5], [0.5, 0.5])
Medium high (MH)	Medium good (MG)	([0.6, 0.7], [0.2, 0.3])
High (H)	Good (G)	([0.7, 0.8], [0.1, 0.2])
Very high (VH)	Very good (VG)	([0.8, 0.9], [0, 0.1])
Extremely high (EH)	Extremely good (EG)	([1, 1], [0, 0])

Step 4: Construct the weighted normalized matrix $\tilde{V} = [\tilde{v}_{ij}]_{m \times n}$.

$$\tilde{v}_{ij} = \tilde{\beta}_{ij} \otimes \tilde{w}_j \tag{7.16}$$

such that $a_{v_{ij}} = a_{\beta_{ij}} \times a_{w_j}$, $b_{v_{ij}} = b_{\beta_{ij}} \times b_{w_j}$, $c_{v_{ij}} = c_{\beta_{ij}} \times c_{w_j}$, $d_{v_{ij}} = d_{\beta_{ij}} \times d_{w_j}$.

Step 5: Determine the ideal solutions.

The positive ideal solution is

$$A^* = \{v_1^*, \ldots, v_m^*\}, \tag{7.17}$$

with $\tilde{v}_j^* = ([1, 1], [0, 0])$ for benefit criteria and $\tilde{v}_j^* = ([0, 0], [1, 1])$ for cost criteria.

The negative ideal solution is

$$A^* = \{v_1^-, \ldots, v_m^-\}, \tag{7.18}$$

with $\tilde{v}_j^- = ([0, 0], [1, 1])$ for benefit criteria and $\tilde{v}_j^- = ([1, 1], [0, 0])$ for cost criteria.

In addition to rank reversal, highlighted in Chapters 4, 5, and 6, the choice of ideal solutions (maximal and minimal reference points) is sometimes confusing in the fuzzy context, since there is no total ordering of fuzzy quantities (El Alaoui & El Yassini, 2020). In the context of intuitionistic fuzzy sets, the higher the membership function is and the lower the nonmembership function is, the higher ranked the intuitionistic fuzzy number should be. Almost all methods presented in the previous section opted to interpret ideal solutions as $\tilde{v}_j^* = ([\max_i a_{v_{ij}}, \max_i b_{v_{ij}}], [\min_i c_{v_{ij}}, \min_i d_{v_{ij}}])$ for benefit criteria and $\tilde{v}_j^* = ([\min_i a_{v_{ij}}, \min_i b_{v_{ij}}], [\max_i c_{v_{ij}}, \max_i d_{v_{ij}}])$ for cost criteria.

Step 6: Compute the distance between each alternative and the ideal solutions.

The distance to the positive ideal solution is

$$D_i^* = \sum_{j=1}^{m} d(\tilde{v}_{ij}, \tilde{v}_j^*), \quad i = 1, \ldots, n, \qquad (7.19)$$

with $d(\tilde{v}_{ij}, \tilde{v}_j^*) = \sqrt{(\frac{1}{4}) * ((a_{v_{ij}} - 1)^2 + (b_{v_{ij}} - 1)^2 + (c_{v_{ij}})^2 + (d_{v_{ij}})^2)}$ for benefit criteria and $d(\tilde{v}_{ij}, \tilde{v}_j^*) = \sqrt{(\frac{1}{4}) * ((a_{v_{ij}})^2 + (b_{v_{ij}})^2 + (c_{v_{ij}} - 1)^2 + (d_{v_{ij}} - 1)^2)}$ for cost criteria.

The distance to the negative ideal solution is

$$D_i^- = \sum_{j=1}^{m} d(\tilde{v}_{ij}, \tilde{v}_j^-), \quad i = 1, \ldots, n, \qquad (7.20)$$

with $d(\tilde{v}_{ij}, \tilde{v}_j^-) = \sqrt{(\frac{1}{4}) * ((a_{v_{ij}})^2 + (b_{v_{ij}})^2 + (c_{v_{ij}} - 1)^2 + (d_{v_{ij}} - 1)^2)}$ for cost criteria and $d(\tilde{v}_{ij}, \tilde{v}_j^-) = \sqrt{(\frac{1}{4}) * ((a_{v_{ij}} - 1)^2 + (b_{v_{ij}} - 1)^2 + (c_{v_{ij}})^2 + (d_{v_{ij}})^2)}$ for benefit criteria.

Step 7: Calculate the relative closeness to the ideal solution:

$$CC_i = \frac{D_i^-}{D_i^- + D_i^*}, \quad i = 1, \ldots, n. \qquad (7.21)$$

Step 8: Rank according to CC_i values, from highest to lowest. ▫

EXAMPLE

Here we use the example presented by Banaeian et al. (2018) and adapted to the interval-valued intuitionistic fuzzy context by Shi et al. (2018) in which four olive-oil suppliers (O1, O2, O3, and O4), three palm-oil suppliers (P1, P2, and P3), and four soybean-oil suppliers (S1, S2, S3, and S4) are evaluated by three decision makers according to four criteria (C_1: service level; C_2: product quality; C_3: price; C_4: environmental management system). Tables 7.10 and 7.11 give, respectively, the individual assessments of alternatives according to each criterion and the individual assessments of criterion importance.

Step 1: Each decision maker assesses each alternative using a linguistic variable $\tilde{\beta}_{ijk}$ (Table 7.10).

Step 2: Each decision maker assesses each criterion's importance using a linguistic variable $\tilde{w}_{jk}$ (Table 7.11).

Step 3: Using Algorithm 7.1, compute the collective alternative assessments and criterion weights (Table 7.12).

Step 4: Construct the weighted normalized matrix (Table 7.13).

Step 5: Determine the ideal solutions.

Since all criteria are benefit criteria, the positive ideal solution is $A^* = \{v_1^*, \ldots, v_{10}^*\}$, with $v_j^* = ([1, 1], [0, 0])$ for $1 \leq j \leq 10$, and the negative ideal solution is $A^- = \{v_1^-, \ldots, v_{10}^-\}$, with $v_j^- = ([0, 0], [1, 1])$ for $1 \leq j \leq 10$.

TABLE 7.10
Alternative Assessment

Supplier Product	Decision Maker	Supplier	Criterion			
			C_1	C_2	C_3	C_4
Olive oil	DM_1	O1	F	MP	MP	VG
		O2	G	F	F	MG
		O3	MP	VG	MG	G
		O4	P	VP	P	MP
	DM_2	O1	P	G	MP	F
		O2	F	G	MG	F
		O3	VG	MP	VG	VG
		O4	MP	MG	F	MG
	DM_3	O1	MP	F	G	MG
		O2	VG	P	G	G
		O3	P	G	G	G
		O4	F	P	G	G
Palm oil	DM_1	P1	G	MG	MG	G
		P2	P	MP	MP	P
		P3	MG	F	F	MG
	DM_2	P1	G	MG	MG	MG
		P2	P	P	P	P
		P3	F	F	G	F
	DM_3	P1	VG	VG	G	G
		P2	MG	F	MP	MP
		P3	MG	F	G	MP
Soybean oil	DM_1	S1	VG	F	VG	F
		S2	MG	MG	MG	MG
		S3	P	VG	F	F
	DM_2	S1	G	MP	G	F
		S2	MP	F	P	F
		S3	P	MP	MP	F
	DM_3	S1	G	G	MG	MG
		S2	MG	MP	VP	VP
		S3	G	P	MG	MG

TABLE 7.11
Criterion Importance

Decision Maker	Criterion			
	C_1	C_2	C_3	C_4
DM_1	H	H	MH	ML
DM_2	VH	MH	H	M
DM_3	H	VH	MH	MH

TABLE 7.12
Consensual Alternative Assessments and Criterion Importance

Alternative	Criterion			
	C_1	C_2	C_3	C_4
O1	([0.28, 0.34], [0.60, 0.66])	([0.48, 0.55], [0.38, 0.45])	([0.38, 0.49], [0.41, 0.51])	([0.7, 0.8], [0.1, 0.2])
O2	([0.68, 0.75], [0.17, 0.25])	([0.46, 0.53], [0.41, 0.47])	([0.6, 0.68], [0.25, 0.32])	([0.6, 0.68], [0.25, 0.32])
O3	([0.40, 0.50], [0.39, 0.50])	([0.6, 0.71], [0.19, 0.29])	([0.7, 0.8], [0.1, 0.2])	([0.73, 0.83], [0.07, 0.17])
O4	([0.28, 0.34], [0.60, 0.66])	([0.25, 0.36], [0.54, 0.64])	([0.46, 0.53], [0.41, 0.47])	([0.52, 0.63], [0.27, 0.37])
P1	([0.73, 0.83], [0.07, 0.17])	([0.67, 0.77], [0.13, 0.23])	([0.63, 0.73], [0.17, 0.27])	([0.67, 0.77], [0.13, 0.23])
P2	([0.28, 0.38], [0.51, 0.62])	([0.28, 0.34], [0.6, 0.66])	([0.17, 0.27], [0.63, 0.73])	([0.13, 0.23], [0.67, 0.77])
P3	([0.57, 0.64], [0.29, 0.36])	([0.5, 0.5], [0.5, 0.5])	([0.64, 0.71], [0.21, 0.29])	([0.45, 0.51], [0.42, 0.49])
S1	([0.73, 0.83], [0.07, 0.17])	([0.48, 0.55], [0.38, 0.45])	([0.7, 0.8], [0.1, 0.2])	([0.53, 0.57], [0.39, 0.43])
S2	([0.49, 0.59], [0.31, 0.41])	([0.45, 0.51], [0.42, 0.49])	([0.25, 0.36], [0.54, 0.64])	([0.4, 0.46], [0.47, 0.54])
S3	([0.33, 0.43], [0.47, 0.57])	([0.40, 0.50], [0.39, 0.50])	([0.45, 0.51], [0.42, 0.49])	([0.53, 0.57], [0.39, 0.43])
Weights	([0.73, 0.83], [0.07, 0.17])	([0.7, 0.8], [0.1, 0.2])	([0.63, 0.73], [0.17, 0.27])	([0.4, 0.51], [0.42, 0.49])

Step 6: Compute the distance between each alternative and the ideal solutions (Table 7.14).
Step 7: Calculate the relative closeness to the ideal solution (Table 7.14).
Step 8: Rank according to CC_i values, from highest to lowest (Table 7.14).

TABLE 7.13
Weighted Normalized Assessments

Alternative	Criterion			
	C_1	C_2	C_3	C_4
O1	([0.2, 0.28], [0.04, 0.11])	([0.34, 0.44], [0.04, 0.09])	([0.24, 0.36], [0.07, 0.14])	([0.31, 0.41], [0.04, 0.1])
O2	([0.5, 0.63], [0.01, 0.04])	([0.32, 0.42], [0.04, 0.09])	([0.38, 0.5], [0.04, 0.09])	([0.27, 0.35], [0.1, 0.16])
O3	([0.29, 0.42], [0.03, 0.08])	([0.43, 0.57], [0.02, 0.06])	([0.45, 0.59], [0.02, 0.05])	([0.33, 0.43], [0.03, 0.08])
O4	([0.2, 0.28], [0.04, 0.11])	([0.18, 0.29], [0.05, 0.13])	([0.29, 0.39], [0.07, 0.13])	([0.23, 0.32], [0.11, 0.18])
P1	([0.54, 0.7], [0, 0.03])	([0.47, 0.62], [0.01, 0.05])	([0.4, 0.54], [0.03, 0.07])	([0.3, 0.39], [0.06, 0.11])
P2	([0.21, 0.32], [0.03, 0.1])	([0.19, 0.27], [0.06, 0.13])	([0.11, 0.2], [0.1, 0.19])	([0.06, 0.12], [0.28, 0.37])
P3	([0.42, 0.53], [0.02, 0.06])	([0.35, 0.4], [0.05, 0.1])	([0.41, 0.52], [0.03, 0.08])	([0.2, 0.26], [0.18, 0.24])
S1	([0.54, 0.7], [0, 0.03])	([0.34, 0.44], [0.04, 0.09])	([0.45, 0.59], [0.02, 0.05])	([0.24, 0.29], [0.17, 0.21])
S2	([0.36, 0.49], [0.02, 0.07])	([0.31, 0.41], [0.04, 0.1])	([0.16, 0.26], [0.09, 0.17])	([0.18, 0.24], [0.2, 0.26])
S3	([0.24, 0.36], [0.03, 0.09])	([0.28, 0.41], [0.04, 0.1])	([0.28, 0.38], [0.07, 0.13])	([0.24, 0.29], [0.17, 0.21])

According to Shi et al. (2018), the ranking obtained here is identical to the ones obtained by fuzzy TOPSIS, fuzzy VIKOR, and fuzzy GRA.

CONTINUOUS CONSENSUAL INTUITIONISTIC FUZZY TOPSIS
THE APPROACH

As explained in Chapter 2, fuzzy logic aims to create a sort of continuity between possible truth values. In that sense, trapezoidal intuitionistic fuzzy sets (Nehi & Maleki, 2005) constitute a continuous generalization of discrete and interval-valued intuitionistic fuzzy sets. The algorithm to derive consensus (Algorithm 6.3) is adapted to trapezoidal intuitionistic fuzzy numbers by modifying the similarity and the distance, respectively, as follows:

$$S(\tilde{R}_k, \tilde{R}) = 1 - \frac{1}{8}(d(\tilde{R}_k, \tilde{R}))^2 \qquad (7.22)$$

TABLE 7.14
Distances to Ideal Solutions, Closeness Coefficients, and Final Ranking

Alternative	D_i^-	D_i^*	CC_i	Rank
O1	2.7763	1.9314	0.5897	3
O2	2.8936	1.6622	0.6351	2
O3	2.9824	1.6085	0.6496	1
O4	2.6616	2.0856	0.5607	4
P1	3.0578	1.4511	0.6782	1
P2	2.4432	2.3684	0.5078	3
P3	2.7968	1.7667	0.6129	2
S1	2.9244	1.5939	0.6472	1
S2	2.6470	2.0195	0.5672	3
S3	2.6850	1.9877	0.5746	2

and

$$d(\tilde{R}_k, \tilde{R}) = (\sum_{q=1}^{4} (|r_k^q - r^q|)^2 + \sum_{q=1}^{4} (|r_k^{q'} - r^{q'}|)^2)^{\frac{1}{2}} \quad (7.23)$$

where $\tilde{R}_k(r_k^1, r_k^2, r_k^3, r_k^4, r_k^{1'}, r_k^{2'}, r_k^{3'}, r_k^{4'})$ is the kth individual opinion and $\tilde{R}(r^1, r^2, r^3, r^4, r^{1'}, r^{2'}, r^{3'}, r^{4'})$ is the aggregated consensus.

An extended TOPSIS algorithm using triangular intuitionistic fuzzy numbers is given by Robinson and Amirtharaj (2011), and one using trapezoidal intuitionistic fuzzy numbers by Li and Chen (2014). The proposed extension here is as follows:

Algorithm 7.3: Let K decision makers ($1 \leq k \leq K$) assess m alternatives ($1 \leq i \leq m$) according to n criteria ($1 \leq j \leq n$).

Step 1: Each decision maker assesses each alternative using a linguistic variable $\tilde{x}_{ijk}$ (Table 7.15).

Step 2: Each decision maker assesses each criterion's importance using a linguistic variable $\tilde{w}_{jk}$ (Table 7.15).

Step 3: Using the individual alternative assessments $\tilde{x}_{ijk}(x_{ijk}^1, x_{ijk}^2, x_{ijk}^3, x_{ijk}^4, x_{ijk}^{1'}, x_{ijk}^{2'}, x_{ijk}^{3'}, x_{ijk}^{4'})$ and criterion weights $\tilde{w}_{jk}(w_{jk}^1, w_{jk}^2, w_{jk}^3, w_{jk}^4, w_{jk}^{1'}, w_{jk}^{2'}, w_{jk}^{3'}, w_{jk}^{4'})$, compute the collective alternative assessments $\tilde{x}_{ij}$ and criterion weights $\tilde{w}_j$ using Algorithm 6.3 while adopting the distance (Equation 7.23) and the similarity (Equation 7.22).

TABLE 7.15
Linguistic Variables for Assessing Alternatives and Criterion Importance using Interval Valued Intuitionistic Fuzzy Numbers

Linguistic Variable for Alternatives	Linguistic Variable for Criteria	Interval-Valued Intuitionistic Fuzzy Number
Very low (VL)	Very poor (VP)	([0, 0, 0, 0], [0, 0, 0, 0])
Low (L)	Poor (P)	([0, 0.1, 0.2, 0.3], [0, 0.1, 0.2, 0.3])
Medium low (ML)	Medium poor (MP)	([0.1, 0.2, 0.3, 0.4], [0, 0.2, 0.3, 0.5])
Medium (M)	Fair (F)	([0.3, 0.4, 0.5, 0.6], [0.2, 0.4, 0.5, 0.7])
Medium high (MH)	Medium good (MG)	([0.5, 0.6, 0.7, 0.8], [0.4, 0.6, 0.7, 0.9])
High (H)	Good (G)	([0.7, 0.8, 0.9, 1], [0.7, 0.8, 0.9, 1])
Very high (VH)	Very good (VG)	([1, 1, 1, 1], [1, 1, 1, 1])

Since all interval-valued intuitionistic fuzzy values used are in the unity interval, no normalization is required.

Step 4: Construct the weighted normalized matrix $\tilde{V} = [\tilde{v}_{ij}]_{m \times n}$.

$$\tilde{v}_{ij} = \tilde{x}_{ij} \otimes \tilde{w}_j \quad (7.24)$$

such that $v_{ij}^q = x_{ij}^q \times w_j^q$ and $v_{ij}^{q'} = v_{ij}^{q'} = x_{ij}^{q'} \times w_j^{q'}$, with $1 \leq q \leq 4$.

Step 5: Determine the ideal solutions.

The positive ideal solution is

$$A^* = \{v_1^*, \ldots, v_m^*\}, \quad (7.25)$$

with $\tilde{v}_j^* = ([1, 1, 1, 1], [1, 1, 1, 1])$ for benefit criteria and $\tilde{v}_j^* = ([0, 0, 0, 0], [0, 0, 0, 0])$ for cost criteria.

The negative ideal solution is

$$A^* = \{v_1^-, \ldots, v_m^-\} \quad (7.26)$$

with $\tilde{v}_j^- = ([0, 0, 0, 0], [0, 0, 0, 0])$ for benefit criteria and $\tilde{v}_j^- = ([1, 1, 1, 1], [1, 1, 1, 1])$ for cost criteria.

Step 6: Compute the distance between each alternative and the ideal solutions.

The distance to the positive ideal solution is

$$D_i^* = \sum_{j=1}^{m} d(\tilde{v}_{ij}, \tilde{v}_j^*), \quad i = 1, \ldots, n, \tag{7.27}$$

with $d(\tilde{v}_{ij}, \tilde{v}_j^*) = \sqrt{(\frac{1}{8}) * \sum_{q=1}^{4}(1-v_{ij}^q)^2 + (1-v_{ij}^{q'})^2}$ for benefit criteria and $d(\tilde{v}_{ij}, \tilde{v}_j^*) = \sqrt{(\frac{1}{8}) * \sum_{q=1}^{4}(v_{ij}^q)^2 + (v_{ij}^{q'})^2}$ for cost criteria.

The distance to the negative ideal solution is

$$D_i^- = \sum_{j=1}^{m} d(\tilde{v}_{ij}, \tilde{v}_j^-), \quad i = 1, \ldots, n, \tag{7.28}$$

with $d(\tilde{v}_{ij}, \tilde{v}_j^-) = \sqrt{(\frac{1}{8}) * \sum_{q=1}^{4}(v_{ij}^q)^2 + (v_{ij}^{q'})^2}$ for benefit criteria and $d(\tilde{v}_{ij}, \tilde{v}_j^-) = \sqrt{(\frac{1}{8}) * \sum_{q=1}^{4}(1-v_{ij}^q)^2 + (1-v_{ij}^{q'})^2}$ for cost criteria.

Step 7: Calculate the relative closeness to the ideal solution:

$$CC_i = \frac{D_i^-}{D_i^- + D_i^*}, \quad i = 1, \ldots, n. \tag{7.29}$$

Step 8: Rank according to CC_i values, from highest to lowest. ☐

EXAMPLE

In an example treated by Li and Chen (2014), four information systems (alternatives A_1, A_2, A_3, and A_4) are evaluated by four decision makers (DM_1, DM_2, DM_3, and DM_4) according to three criteria (C_1: cost; C_2: contribution to organizational performance improvement; C_3: reliability). Tables 7.16 and 7.17 detail the linguistic assessments for alternatives and criterion weights.

Step 1: Each decision maker assesses each alternative using a linguistic variable $\tilde{x}_{ijk}$ (Table 7.16).

Step 2: Each decision maker assesses each criterion's importance using a linguistic variable $\tilde{w}_{jk}$ (Table 7.17).

Step 3: Compute the collective alternative assessments and criterion weights using Algorithm 6.3 and Equations 7.22 and 7.23 (Table 7.18).

Step 4: Construct the weighted normalized matrix $\tilde{V} = [\tilde{v}_{ij}]_{m \times n}$ (Table 7.19).

Step 5: Determine the ideal solutions.

Since C_1 is a cost criterion and C_2 and C_3 are benefit criteria, the positive ideal solution is

$$A^* = \{([0, 0, 0, 0], [0, 0, 0, 0]), ([1, 1, 1, 1], [1, 1, 1, 1]),$$
$$([1, 1, 1, 1], [1, 1, 1, 1]),$$

TABLE 7.16
Linguistic Variables for Assessing Alternatives

Decision Maker	Alternative	Criterion		
		C_1	C_2	C_3
DM_1	A_1	ML	L	VH
	A_2	VH	H	ML
	A_3	M	MH	ML
	A_4	MH	M	MH
DM_2	A_1	M	ML	H
	A_2	H	VH	M
	A_3	L	MH	ML
	A_4	H	MH	M
DM_3	A_1	MH	ML	H
	A_2	MH	H	M
	A_3	L	M	L
	A_4	M	M	MH
DM_4	A_1	M	L	H
	A_2	H	VH	M
	A_3	ML	M	M
	A_4	MH	M	M

TABLE 7.17
Linguistic Variables for Criterion Importance

Decision Maker	Criterion		
	C_1	C_2	C_3
DM_1	F	F	F
DM_2	MG	MP	MG
DM_3	MG	F	P
DM_4	MG	F	F

and the negative ideal solution is

$$A^* = \{([1, 1, 1, 1], [1, 1, 1, 1]), ([0, 0, 0, 0], [0, 0, 0, 0]), ([0, 0, 0, 0], [0, 0, 0, 0])\}.$$

Step 6: Compute the distance between each alternative and the ideal solutions (Table 7.20).

TABLE 7.18
Consensual Alternative and Criterion Assessments

Aggregated Assessment	Criterion		
	C_1	C_2	C_3
A_1	([0.30, 0.40, 0.50, 0.60], [0.20, 0.40, 0.50, 0.70])	([0.05, 0.15, 0.25, 0.35], [0, 0.15, 0.25, 0.4])	([0.77, 0.85, 0.92, 1], [0.77, 0.85, 0.92, 1])
A_2	([0.72, 0.80, 0.88, 0.95], [0.70, 0.80, 0.88, 0.98])	([0.85, 0.9, 0.95, 1], [0.85, 0.9, 0.95, 1])	([0.25, 0.35, 0.45, 0.55], [0.15, 0.35, 0.45, 0.65])
A_3	([0.10, 0.20, 0.30, 0.40], [0.05, 0.20, 0.30, 0.45])	([0.40, 0.50, 0.60, 0.70], [0.30, 0.50, 0.60, 0.80])	([0.12, 0.22, 0.32, 0.42], [0.05, 0.22, 0.32, 0.5])
A_4	([0.50, 0.60, 0.70, 0.80], [0.42, 0.60, 0.70, 0.88])	([0.35, 0.45, 0.55, 0.65], [0.25, 0.45, 0.55, 0.75])	([0.40, 0.50, 0.60, 0.70], [0.30, 0.50, 0.60, 0.80])
Weights	([0.54, 0.55, 0.65, 0.75], [0.35, 0.55, 0.65, 0.85])	([0.25, 0.35, 0.45, 0.55], [0.15, 0.35, 0.45, 0.65])	([0.28, 0.38, 0.48, 0.58], [0.20, 0.38, 0.48, 0.65])

TABLE 7.19
Weighted Normalized Decision Matrix

Alternative	Criterion		
	C_1	C_2	C_3
A_1	([0.14, 0.22, 0.33, 0.45], [0.07, 0.22, 0.33, 0.60])	([0.01, 0.05, 0.11, 0.19], [0, 0.05, 0.11, 0.26])	([0.21, 0.32, 0.44, 0.58], [0.16, 0.32, 0.44, 0.65])
A_2	([0.33, 0.44, 0.57, 0.71], [0.25, 0.44, 0.57, 0.83])	([0.21, 0.32, 0.43, 0.55], [0.13, 0.32, 0.43, 0.65])	([0.07, 0.13, 0.21, 0.32], [0.03, 0.13, 0.21, 0.42])
A_3	([0.04, 0.11, 0.19, 0.30], [0.02, 0.11, 0.19, 0.38])	([0.10, 0.18, 0.27, 0.39], [0.05, 0.18, 0.27, 0.52])	([0.03, 0.08, 0.15, 0.24], [0.01, 0.08, 0.15, 0.33])
A_4	([0.23, 0.33, 0.46, 0.60], [0.15, 0.33, 0.46, 0.74])	([0.09, 0.16, 0.25, 0.36], [0.04, 0.16, 0.25, 0.49])	([0.11, 0.19, 0.29, 0.40], [0.06, 0.19, 0.29, 0.52])

Step 7: Calculate the relative closeness to the ideal solution (Table 7.20).

Step 8: Rank according to CC_i values, from highest to lowest (Table 7.20).

The differences between the ranking proposed here and the one given by Li and Chen (2014) can be explained mainly by the expected value of their trapezoidal intuitionistic fuzzy numbers—a sort of early defuzzification after computing the

TABLE 7.20
Distances to Ideal Solutions, Closeness Coefficients, and Proposed and Compared Rankings

Alternative	D_i^+	D_i^-	CC_i	Proposed Rank	Rank (Li & Chen 2014)
A_1	0.4073	0.3000	0.4242	2	1
A_2	0.4154	0.2468	0.3727	4	3
A_3	0.4171	0.3187	0.4331	1	2
A_4	0.4184	0.2584	0.3818	3	4

weighted normalized matrix, which results in information loss. In the example here, the information is kept until the distances to the ideal solutions are computed.

The interpretation of ideal solutions cannot explain the discordances between rankings in this situation, because using the classical reference points extended for trapezoidal intuitionistic fuzzy numbers by Li and Chen (2014),

$\tilde{v}_j^* = ([\max_i v_{ij}^q], [\max_i v_{ij}^{q'}])$ and $\tilde{v}_j^- = ([\min_i v_{ij}^q], [\min_i v_{ij}^{q'}])$ with $1 \leq q \leq 4$ for benefit criteria

And $\tilde{v}_j^* = ([\min_i v_{ij}^q], [\min_i v_{ij}^{q'}])$ and $\tilde{v}_j^- = ([\max_i v_{ij}^q], [\max_i v_{ij}^{q'}])$ with $1 \leq q \leq 4$ for cost criteria

results in the same final ranking (Table 7.21).

UNKNOWN ATTRIBUTE WEIGHTS

In all the methods mentioned so far, it is supposed that the decision makers can assess criterion weights. Without that assumption, a series of entropy methods have been proposed in the literature (Ding et al., 2016; Freeman & Chen, 2015; Li et al. 2011; Zhang et al., 2011) and adapted to fuzzy sets (Cavallaro, Zavadskas, & Raslanas, 2016; Collan, Fedrizzi, & Luukka, 2015; Chaghooshi, Fathi, & Kashef, 2012; Mavi, Goh, & Mavi, 2016; Reddy, Kumar, & Raj, 2019; Wang, Lee, & Chang, 2007; Won, Chung, & Choi, 2015; Xin, 2010), Pythagorean fuzzy sets (Biswas & Sarkar, 2019; Han et al., 2019, 2020; Lin, Huang, & Xu, 2019; introduced by Yager, 2013), hesitant fuzzy sets (Hussain & Yang, 2018; introduced by Torra 2010), type 2 fuzzy sets (Zamri & Abdullah, 2014), intuitionistic fuzzy sets (Cavallaro et al., 2019; Chen, 2019; Ding & Wang, 2019; Joshi & Kumar, 2014; Pu, Zeng, & Li, 2013; Sachdeva & Kapur 2019; Wood, 2016), and interval-valued intuitionistic fuzzy sets (Zhang & Yu, 2012), whether criterion weights are completely unknown (Biswas & Sarkar, 2019; Chen, Liu, & Tang, 2008; Li et al., 2017; Liu & Du, 2008) or incomplete (Khan et al., 2018, 2020; Liu & Zhang, 2014; Wei, 2010; Xu & Zhang, 2013; Yang & Peng, 2017).

TABLE 7.21
Distances to Max-Min Ideal Solutions, Closeness Coefficients, and Final Ranking

Alternative	D_i^+	D_i^-	CC_i	Proposed Rank
A_1	0.4214	0.4864	0.5358	2
A_2	0.5565	0.3495	0.3858	4
A_3	0.3989	0.5098	0.5610	1
A_4	0.5468	0.3667	0.4014	3

THE APPROACH

To permit a focus on the process of obtaining criterion weights, Shen et al. (2018) have proposed a relatively simple approach in the intuitionistic fuzzy context:

Step 1: Establish the intuitionistic fuzzy decision matrix.

In this step, m alternatives ($1 \leq i \leq m$) are evaluated according to n criteria ($1 \leq j \leq n$) using intuitionistic fuzzy numbers $\alpha_{ij} = (\mu_{ij}, \vartheta_{ij})$.

Step 2: Determine the ideal solutions.

While Shen et al. (2018) opted for the classical interpretation of ideal solutions—$\alpha_j^*(\max_i \mu_{ij}, \min_i \vartheta_{ij})$ and $\alpha_j^-(\min_i \mu_{ij}, \max_i \vartheta_{ij})$ for benefit criteria and $\alpha_j^*(\min_i \mu_{ij}, \max_i \vartheta_{ij})$ and $\alpha_j^-(\max_i \mu_{ij}, \min_i \vartheta_{ij})$ for cost criteria—here, to prevent rank reversal (El Alaoui, El Yassini, & Ben-azza, 2019), the positive ideal solution is

$$A^* = \{\alpha_1^*, \ldots, \alpha_n^*\}, \quad (7.30)$$

with $\alpha_j^* = (1, 0)$ for benefit criteria and $\alpha_j^- = (0, 1)$ for cost criteria, and the negative ideal solution is

$$A^- = \{\alpha_1^-, \ldots, \alpha_n^-\}, \quad (7.31)$$

with $\alpha_j^* = (0, 1)$ for benefit criteria and $\alpha_j^- = (1, 0)$ for cost criteria.

Step 3: Compute the distances between each alternative according to each criterion α_{ij} and the ideal solutions $\alpha_j^*(\mu_j^*, \vartheta_j^*)$ and $\alpha_j^-(\mu_j^-, \vartheta_j^-)$.

Hence the distances to the positive ideal solutions are

$$d_{ij}^* = \sqrt{0.5 * ((\mu_{ij} - 1)^2 + (\vartheta_{ij})^2)} \quad (7.32)$$

and the distances to the negative ideal solutions are

$$d_{ij}^- = \sqrt{0.5 * ((\mu_{ij})^2 + (\vartheta_{ij} - 1)^2)}. \tag{7.33}$$

Step 4: Calculate the composite intuitionistic distances Z_{ij}:

$$Z_{ij} = d_{ij}^- - d_{ij}^*. \tag{7.34}$$

Step 5: Compute the criterion weights w_j using

$$w_j = \frac{\sum_{i=1}^{m} \sum_{o=1}^{m} |Z_{ij} - Z_{oj}|}{\sum_{j=1}^{n} \sum_{i=1}^{m} \sum_{o=1}^{m} |Z_{ij} - Z_{oj}|}. \tag{7.35}$$

Step 6: Compute the weighted intuitionistic distances

$$D_i = \sum_{j=1}^{n} w_j * Z_{ij}, \ i = 1, \ldots, m. \tag{7.36}$$

The weighted distance plays the role of the closeness coefficient.

Step 7: Rank alternatives according to D_i values, from highest to lowest. ▫

EXAMPLE

In this example from Shen et al. (2018), a financial enterprise desires to select a strategic partner from among five companies (alternatives A_1, A_2, A_3, A_4, and A_5) according to five criteria (C_1: character; C_2: capacity; C_3: capital; C_4: collateral; C_5: condition).

Step 1: Establish the intuitionistic fuzzy decision matrix

Suppose the evaluation matrix is given in Table 7.22 (Shen et al. 2018).

Step 2: Determine the ideal solutions.

Since all criteria are benefit criteria, the positive ideal solution is

$$A^* = \{(1, 0), (1, 0), (1, 0), (1, 0), (1, 0)\}$$

and the negative ideal solution is

$$A^- = \{(0, 1), (0, 1), (0, 1), (0, 1), (0, 1)\}.$$

Step 3: Compute the distances to the ideal solutions (Table 7.23).
Step 4: Calculate the composite intuitionistic distances Z_{ij} (Table 7.24).
Step 5: Compute the criterion weights:

$w_1 = 0.0999$, $w_2 = 0.2628$, $w_3 = 0.2474$, $w_4 = 0.2258$ and $w_5 = 0.1641$.

Step 6: Compute the weighted intuitionistic distances:

$D_1 = 0.1554$, $D_2 = 0.3265$, $D_3 = 0.2926$, $D_4 = -0.107$ and $D_5 = 0.1114$.

Step 7: Rank alternatives according to D_i values, from highest to lowest.

TABLE 7.22
Evaluation Matrix of Five Alternatives According to Five Criteria

Alternative	Criterion				
	C_1	C_2	C_3	C_4	C_5
A_1	(0.6, 0.3)	(0.4, 0.3)	(0.3, 0.5)	(0.6, 0.3)	(0.7, 0.2)
A_2	(0.6, 0.2)	(0.5, 0.3)	(0.5, 0.1)	(0.6, 0.3)	(0.6, 0.1)
A_3	(0.8, 0.1)	(0.3, 0.2)	(0.6, 0.1)	(0.5, 0.4)	(0.5, 0.1)
A_4	(0.7, 0.2)	(0.1, 0.6)	(0.5, 0.3)	(0.2, 0.5)	(0.5, 0.6)
A_5	(0.6, 0.2)	(0.5, 0.1)	(0.3, 0.4)	(0.2, 0.5)	(0.6, 0.2)

TABLE 7.23
Distances to Ideal Solutions

Alternative	d_{ij}^*					d_{ij}^-				
	C_1	C_2	C_3	C_4	C_5	C_1	C_2	C_3	C_4	C_5
A_1	0.3536	0.4743	0.6083	0.3536	0.2550	0.6519	0.5701	0.4123	0.6519	0.7517
A_2	0.3162	0.4123	0.3606	0.3536	0.2915	0.7071	0.6083	0.7280	0.6519	0.7649
A_3	0.1581	0.5148	0.2915	0.4528	0.3606	0.8515	0.6042	0.7649	0.5523	0.7280
A_4	0.2550	0.7649	0.4123	0.6671	0.5523	0.7517	0.2915	0.6083	0.3808	0.4528
A_5	0.3162	0.3606	0.5701	0.6671	0.3162	0.7071	0.7280	0.4743	0.3808	0.7071

TABLE 7.24
Composite Intuitionistic Distances

Alternative	C_1	C_2	C_3	C_4	C_5
A_1	0.2984	0.0957	−0.1960	0.2984	0.4967
A_2	0.3909	0.1960	0.3675	0.2984	0.4733
A_3	0.6934	0.0894	0.4733	0.0995	0.3675
A_4	0.4967	−0.4733	0.1960	−0.2863	−0.0995
A_5	0.3909	0.3675	−0.0957	−0.2863	0.3909

TABLE 7.25
Proposed Ranking and Ranking Obtained by Shen et al. (2018)

Alternative	Proposed Ranking	Ranking of Shen et al. (2018)
A_1	3	4
A_2	1	1
A_3	2	2
A_4	5	5
A_5	4	3

Table 7.25 details the ranking proposed here and the one obtained by Shen et al. (2018). Although they are similar—within one permutation of each other, between A_1 and A_5 ranked third and fourth—it illustrates the influence of ideal solutions on the final ranking.

CONCLUSION

This chapter is devoted to intuitionistic fuzzy TOPSIS extensions. It sets the aggregation (fusion) of several decision makers' opinions as part of the process of information integration, with a detailed presentation of steps and a classification of aggregation functions. Then the chapter discusses the classically used aggregation functions and distances in the intuitionistic fuzzy context, exposes their drawbacks, and presents improved measures. Then, based on the measures introduced, it proposes a fuzzy intuitionistic consensus-reaching algorithm that permits faster convergence. After that, the algorithm is incorporated into the TOPSIS framework and compared to other intuitionistic fuzzy TOPSIS approaches with an example. The chapter also discusses the extension to continuous intuitionistic fuzzy sets and the case of unknown attribute weights.

REFERENCES

Atanassov, Krassimir T. (1999). *Intuitionistic Fuzzy Sets: Theory and Applications*. Studies in Fuzziness and Soft Computing 35. Physica-Verlag Heidelberg. //www.springer.com/gp/book/9783790812282.

Banaeian, Narges, Hossein Mobli, Behnam Fahimnia, Izabela Ewa Nielsen, & Mahmoud Omid (2018). "Green Supplier Selection Using Fuzzy Group Decision Making Methods: A Case Study from the Agri-Food Industry." *Computers & Operations Research*, 89(January), 337–347. https://doi.org/10.1016/j.cor.2016.02.015.

Beliakov, Gleb, Ana Pradera, & Tomasa Calvo (2007). *Aggregation Functions: A Guide for Practitioners*. Berlin: Springer.

Biswas, Animesh, & Biswajit Sarkar (2019). "Pythagorean Fuzzy TOPSIS for Multicriteria Group Decision-Making with Unknown Weight Information through Entropy Measure: BISWAS AND SARKAR." *International Journal of Intelligent Systems*, 34(6), 1108–1128. https://doi.org/10.1002/int.22088.

Cavallaro, Fausto, Edmundas Kazimieras Zavadskas, Dalia Streimikiene, & Abbas Mardani (2019). "Assessment of Concentrated Solar Power (CSP) Technologies Based on a

Modified Intuitionistic Fuzzy Topsis and Trigonometric Entropy Weights." *Technological Forecasting and Social Change*, *140*(March), 258–270. https://doi.org/10.1016/j.techfore.2018.12.009.

Cavallaro, Fausto, Edmundas Zavadskas, & Saulius Raslanas (2016). "Evaluation of Combined Heat and Power (CHP) Systems Using Fuzzy Shannon Entropy and Fuzzy TOPSIS." *Sustainability*, *8*(6), 556. https://doi.org/10.3390/su8060556.

Chen (2019). "A New Multi-Criteria Assessment Model Combining GRA Techniques with Intuitionistic Fuzzy Entropy-Based TOPSIS Method for Sustainable Building Materials Supplier Selection." *Sustainability*, *11*(8), 2265. https://doi.org/10.3390/su11082265.

Chen, Rong, Peide Liu, & Shukun Tang (2008). "Research on Supply-Chain Supplier Selection Based on TOPSIS Method with Interval Number and Unknown Weight." In 2008 4th International Conference on Wireless Communications, Networking and Mobile Computing, 1–4. Dalian, China: IEEE. https://doi.org/10.1109/WiCom.2008.1655.

Chu, Junfeng, Xinwang Liu, Yingming Wang, & Kwai-Sang Chin (2016). "A Group Decision Making Model Considering Both the Additive Consistency and Group Consensus of Intuitionistic Fuzzy Preference Relations." *Computers & Industrial Engineering*, *101*(November), 227–242. https://doi.org/10.1016/j.cie.2016.08.018.

Collan, Mikael, Mario Fedrizzi, & Pasi Luukka (2015). "New Closeness Coefficients for Fuzzy Similarity Based Fuzzy TOPSIS: An Approach Combining Fuzzy Entropy and Multidistance." *Advances in Fuzzy Systems*, *2015*, 1–12. https://doi.org/10.1155/2015/251646.

De, Avijit, Sujit Das, & Samarjit Kar (2019). "Multiple Attribute Decision Making Based on Probabilistic Interval-Valued Intuitionistic Hesitant Fuzzy Set and Extended TOPSIS Method." *Journal of Intelligent & Fuzzy Systems*, *37*(4), 5229–5248. https://doi.org/10.3233/JIFS-190205.

Deza, Michel-Marie, & Elena Deza (2006). *Dictionary of Distances*. Netherlands: Elsevier.

Ding, Lin, Zhenfeng Shao, Hanchao Zhang, Cong Xu, and Dewen Wu (2016). "A Comprehensive Evaluation of Urban Sustainable Development in China Based on the TOPSIS-Entropy Method." *Sustainability*, *8*(8), 746. https://doi.org/10.3390/su8080746.

Ding, Quanyu, & Ying-Ming Wang (2019). "Intuitionistic Fuzzy TOPSIS Multi-Attribute Decision Making Method Based on Revised Scoring Function and Entropy Weight Method." *Journal of Intelligent & Fuzzy Systems*, *36*(1), 625–635. https://doi.org/10.3233/JIFS-18963.

El Alaoui, Mohamed, & Hussain Ben-azza (2017). "Generalization of the Weighted Product Aggregation Applied to Data Fusion of Intuitionistic Fuzzy Quantities." In *2017 Intelligent Systems and Computer Vision (ISCV)*, 1–6. https://doi.org/10.1109/ISACV.2017.8054908.

El Alaoui, Mohamed, Hussain Ben-azza, & Khalid El Yassini (2018). "Optimal Weighting Method for Interval-Valued Intuitionistic Fuzzy Opinions." *Notes on Intuitionistic Fuzzy Sets*, *24*(3), 106–110. https://doi.org/10.7546/nifs.2018.24.3.106-110.

El Alaoui, Mohamed, Hussain Ben-azza, & Khalid El Yassini (2019). "Achieving Consensus in Interval Valued Intuitionistic Fuzzy Environment." *Procedia Computer Science*, THE SECOND INTERNATIONAL CONFERENCE ON INTELLIGENT COMPUTING IN DATA SCIENCES, ICDS2018, *148*(January), 218–225. https://doi.org/10.1016/j.procs.2019.01.064.

El Alaoui, Mohamed, & Khalid El Yassini (2020). "Fuzzy Similarity Relations in Decision Making." *Handbook of Research on Emerging Applications of Fuzzy Algebraic Structures*, 369–385. https://doi.org/10.4018/978-1-7998-0190-0.ch020.

El Alaoui, Mohamed, Khalid El Yassini, & Hussain Ben-azza (2019). "Type 2 Fuzzy TOPSIS for Agriculture MCDM Problems." *International Journal of Sustainable Agricultural Management and Informatics*, *5*(2/3), 112–130. https://doi.org/10.1504/IJSAMI.2019.101672.

Freeman, James, & Tao Chen (2015). "Green Supplier Selection Using an AHP-Entropy-TOPSIS Framework." Edited by Dr. Gary Graham. *Supply Chain Management: An International Journal*, 20(3), 327–340. https://doi.org/10.1108/SCM-04-2014-0142.

Gleb, Beliakov, Bustince Sola Humberto, & Calvo Sánchez Tomasa (2016). *A Practical Guide to Averaging Functions*. Switzerland: Springer International Publishing. http://www.springer.com/us/book/9783319247519.

Grabisch, Michel, Jean-Luc Marichal, Radko Mesiar, & Endre Pap (2009). *Aggregation Functions*. United Kingdom: Cambridge University Press.

Gupta, Pankaj, Mukesh Kumar Mehlawat, Nishtha Grover, & Witold Pedrycz (2018). "Multi-Attribute Group Decision Making Based on Extended TOPSIS Method under Interval-Valued Intuitionistic Fuzzy Environment." *Applied Soft Computing*, 69(August), 554–567. https://doi.org/10.1016/j.asoc.2018.04.032.

Han, Li, Song, Zhang, and Wang (2019). "A New Method for MAGDM Based on Improved TOPSIS and a Novel Pythagorean Fuzzy Soft Entropy." *Symmetry*, 11(7), 905. https://doi.org/10.3390/sym11070905.

Han, Qi, Weimin Li, Yanli Lu, Mingfa Zheng, Wen Quan, & Yafei Song (2020). "TOPSIS Method Based on Novel Entropy and Distance Measure for Linguistic Pythagorean Fuzzy Sets With Their Application in Multiple Attribute Decision Making." *IEEE Access*, 8, 14401–14412. https://doi.org/10.1109/ACCESS.2019.2963261.

Hussain, Zahid, & Miin-Shen Yang (2018). "Entropy for Hesitant Fuzzy Sets Based on Hausdorff Metric with Construction of Hesitant Fuzzy TOPSIS." *International Journal of Fuzzy Systems*, 20(8), 2517–2533. https://doi.org/10.1007/s40815-018-0523-2.

Jafarnejad Chaghooshi, Fathi, Ahmad Mohammad Reza, & Mojtaba Kashef (2012). "Integration of Fuzzy Shannon's Entropy with Fuzzy TOPSIS for Industrial Robotic System Section." *Journal of Industrial Engineering and Management*, 5(1), 102–114. https://doi.org/10.3926/jiem.397.

Jiang, Yuan, Zeshui Xu, & Xiaohan Yu (2013). "Compatibility Measures and Consensus Models for Group Decision Making with Intuitionistic Multiplicative Preference Relations." *Applied Soft Computing*, 13(4), 2075–2086. https://doi.org/10.1016/j.asoc.2012.11.007.

John Robinson, P., & E. C. Henry Amirtharaj (2011). "Extended TOPSIS with Correlation Coefficient of Triangular Intuitionistic Fuzzy Sets for Multiple Attribute Group Decision Making." *IJDSST*, 3(3), 15–41. https://doi.org/10.4018/jdsst.2011070102.

Joshi, Deepa, & Sanjay Kumar (2014). "Intuitionistic Fuzzy Entropy and Distance Measure Based TOPSIS Method for Multi-Criteria Decision Making." *Egyptian Informatics Journal*, 15(2), 97–104. https://doi.org/10.1016/j.eij.2014.03.002.

Khan, Muhammad Sajjad Ali, Faisal Khan, Joseph Lemley, Saleem Abdullah, & Fawad Hussain (2020). "Extended Topsis Method Based on Pythagorean Cubic Fuzzy Multi-Criteria Decision Making with Incomplete Weight Information." *Journal of Intelligent & Fuzzy Systems*, 38(2), 2285–2296. https://doi.org/10.3233/JIFS-191089.

Li, Deng-Feng (2010). "TOPSIS-Based Nonlinear-Programming Methodology for Multiattribute Decision Making With Interval-Valued Intuitionistic Fuzzy Sets." *IEEE Transactions on Fuzzy Systems*, 18(2), 299–311. https://doi.org/10.1109/TFUZZ.2010.2041009.

Li, Xiangxin, Kongsen Wang, Liwen Liu, Jing Xin, Hongrui Yang, & Chengyao Gao (2011). "Application of the Entropy Weight and TOPSIS Method in Safety Evaluation of Coal Mines." *Procedia Engineering*, ISMSSE2011, 26(January), 2085–2091. https://doi.org/10.1016/j.proeng.2011.11.2410.

Li, Xihua, & Xiaohong Chen (2014). "Extension of the TOPSIS Method Based on Prospect Theory and Trapezoidal Intuitionistic Fuzzy Numbers for Group Decision Making." *Journal of Systems Science and Systems Engineering*, 23(2), 231–247. https://doi.org/10.1007/s11518-014-5244-y.

Li, Yan-Lai, Rui Wang, & Kwai-Sang Chin (2019). "New Failure Mode and Effect Analysis Approach Considering Consensus under Interval-Valued Intuitionistic Fuzzy Environment." *Soft Computing*, January. https://doi.org/10.1007/s00500-018-03706-5.

Li, Yupeng, Xiaozhen Lian, Cheng Lu, & Zhaotong Wang (2017). "A Large Group Decision Making Approach Based on TOPSIS Framework with Unknown Weights Information." Edited byL. Zhao, A. Xavior, J. Cai, & L. You. *MATEC Web of Conferences* 100, 02013. https://doi.org/10.1051/matecconf/201710002013.

Liao, Huchang, Zeshui Xu, Xiao-Jun Zeng, & Dong-Ling Xu (2016). "An Enhanced Consensus Reaching Process in Group Decision Making with Intuitionistic Fuzzy Preference Relations." *Information Sciences*, Special issue on Discovery Science, *329*(February), 274–286. https://doi.org/10.1016/j.ins.2015.09.024.

Lin, Mingwei, Chao Huang, & Zeshui Xu (2019). "TOPSIS Method Based on Correlation Coefficient and Entropy Measure for Linguistic Pythagorean Fuzzy Sets and Its Application to Multiple Attribute Decision Making." *Complexity*, *2019*(October), 1–16. https://doi.org/10.1155/2019/6967390.

Liu, Fang, & Wei-Guo Zhang (2014). "TOPSIS-Based Consensus Model for Group Decision-Making With Incomplete Interval Fuzzy Preference Relations." *IEEE Transactions on Cybernetics*, *44*(8), 1283–1294. https://doi.org/10.1109/TCYB.2013.2282037.

Liu, Peide, and Zhengwei Du (2008). "Application of E-Commerce Risk Assessment Research with Weight Unknown TOPSIS Method." In 2008 International Symposiums on Information Processing, 345–349. Moscow, Russia: IEEE. https://doi.org/10.1109/ISIP.2008.98.

Mavi, Reza Kiani, Mark Goh, & Neda Kiani Mavi (2016). "Supplier Selection with Shannon Entropy and Fuzzy TOPSIS in the Context of Supply Chain Risk Management." *Procedia - Social and Behavioral Sciences*, *235*(November), 216–225. https://doi.org/10.1016/j.sbspro.2016.11.017.

Meng, Fanyong, & Jie Tang (2013). "Interval-Valued Intuitionistic Fuzzy Multiattribute Group Decision Making Based on Cross Entropy Measure and Choquet Integral." *International Journal of Intelligent Systems*, *28*(12), 1172–1195. https://doi.org/10.1002/int.21624.

Nehi, Hassan Mishmast, & Hamid Reza Maleki (2005). "Intuitionistic Fuzzy Numbers and It's Applications in Fuzzy Optimization Problem." In *Proceedings of the 9th WSEAS International Conference on Systems*, 1–5. ICS'05. Athens, Greece: World Scientific and Engineering Academy and Society (WSEAS).

Park, Dong Gun, Young Chel Kwun, Jin Han Park, & Il Young Park (2009). "Correlation Coefficient of Interval-Valued Intuitionistic Fuzzy Sets and Its Application to Multiple Attribute Group Decision Making Problems." *Mathematical and Computer Modelling*, *50*(9), 1279–1293. https://doi.org/10.1016/j.mcm.2009.06.010.

Park, Jin Han, Il Young Park, Young Chel Kwun, & Xuegong Tan (2011). "Extension of the TOPSIS Method for Decision Making Problems under Interval-Valued Intuitionistic Fuzzy Environment." *Applied Mathematical Modelling*, *35*(5), 2544–2556. https://doi.org/10.1016/j.apm.2010.11.025.

Pu, Hong Bin, Xi Nan Zeng, & Wei Guang Li (2013). "Configuration Evaluation of Printing Machine Based on Intuitionistic Fuzzy Entropy and TOPSIS." *Advanced Materials Research*, *646*(January), 113–119. https://doi.org/10.4028/www.scientific.net/AMR.646.113.

Reddy, A. Suchith, P. Rathish Kumar, & P. Anand Raj (2019). "Entropy-Based Fuzzy TOPSIS Framework for Selection of a Sustainable Building Material." *International Journal of Construction Management*, November, 1–12. https://doi.org/10.1080/15623599.2019.1683695.

Sachdeva, Nitin, & P. K. Kapur (2019). "A Hybrid Intuitionistic Fuzzy and Entropy Weight Based Multi-Criteria Decision Model with TOPSIS." In *System Performance and Management Analytics*, edited byP. K. Kapur, Yury Klochkov, Ajit Kumar Verma, &

Gurinder Singh, 333–345. Asset Analytics. Singapore: Springer Singapore. https://doi.org/10.1007/978-981-10-7323-6_27.

Sajjad Ali Khan, Muhammad, Asad Ali, Saleem Abdullah, Fazli Amin, & Fawad Hussain (2018). "New Extension of TOPSIS Method Based on Pythagorean Hesitant Fuzzy Sets with Incomplete Weight Information." *Journal of Intelligent & Fuzzy Systems*, *35*(5), 5435–5448. https://doi.org/10.3233/JIFS-171190.

Salih, Mahmood M., B. B. Zaidan, A. A. Zaidan, & Mohamed A. Ahmed (2019). "Survey on Fuzzy TOPSIS State-of-the-Art between 2007 and 2017." *Computers & Operations Research*, *104*(April), 207–227. https://doi.org/10.1016/j.cor.2018.12.019.

Santini, Simone, & Ramesh Jain (1997). "Similarity Is a Geometer." *Multimedia Tools Appl*, *5*(3), 277–306. https://doi.org/10.1023/A:1009651725256.

Shen, Feng, Xinsong Ma, Zhiyong Li, Zeshui Xu, & Dongliang Cai (2018). "An Extended Intuitionistic Fuzzy TOPSIS Method Based on a New Distance Measure with an Application to Credit Risk Evaluation." *Information Sciences*, *428*(February), 105–119. https://doi.org/10.1016/j.ins.2017.10.045.

Shi, Hua, Mei-Yun Quan, Hu-Chen Liu, Chun-Yan Duan, Hua Shi, Mei-Yun Quan, Hu-Chen Liu, & Chun-Yan Duan (2018). "A Novel Integrated Approach for Green Supplier Selection with Interval-Valued Intuitionistic Uncertain Linguistic Information: A Case Study in the Agri-Food Industry." *Sustainability*, *10*(3), 733. https://doi.org/10.3390/su10030733.

Solairaju, A., P. John Robinson, & S. Rethinakumar (2014). "IJMTT - Interval Valued Intuitionistic Fuzzy MAGDM Problems with OWA Entropy Weights." *International Journal of Mathematics Trends and Technology IJMTT*, *9*(2), 153–158.

Szmidt, Eulalia (2014). *Distances and Similarities in Intuitionistic Fuzzy Sets*. Studies in Fuzziness and Soft Computing. Switzerland: Springer International Publishing. www.springer.com/la/book/9783319016399.

Szmidt, Eulalia, & Janusz Kacprzyk (2003). "A Consensus-Reaching Process under Intuitionistic Fuzzy Preference Relations." *International Journal of Intelligent Systems*, *18*(7), 837–852. https://doi.org/10.1002/int.10119.

Tan, Chunqiao, & Qiang Zhang (2006). "Fuzzy Multiple Attribute Decision Making Based on Interval Valued Intuitionistic Fuzzy Sets." In *2006 IEEE International Conference on Systems, Man and Cybernetics*, *2*, 1404–1407. https://doi.org/10.1109/ICSMC.2006.384913.

Tien-Chin Wang, Hsien-Da Lee, & Michael Chao-Sheng Chang (2007). "A Fuzzy TOPSIS Approach with Entropy Measure for Decision-Making Problem." In *2007 IEEE International Conference on Industrial Engineering and Engineering Management*, 124–128. Singapore: IEEE. https://doi.org/10.1109/IEEM.2007.4419164.

Torra, Vicenç (2010). "Hesitant Fuzzy Sets." *International Journal of Intelligent Systems*, *25*(6), 529–539. https://doi.org/10.1002/int.20418.

Torra, Vicenç, & Yasuo Narukawa (2007). *Modeling Decisions: Information Fusion and Aggregation Operators*. Cognitive Technologies. Berlin, Heidelberg: Springer-Verlag. //www.springer.com/us/book/9783540687894.

Wan, Shuping, & Jiuying Dong (2014). "A Possibility Degree Method for Interval-Valued Intuitionistic Fuzzy Multi-Attribute Group Decision Making." *Journal of Computer and System Sciences*, *80*(1), 237–256. https://doi.org/10.1016/j.jcss.2013.07.007.

Wan, Shuping, Feng Wang, & Jiuying Dong (2018). "A Group Decision-Making Method Considering Both the Group Consensus and Multiplicative Consistency of Interval-Valued Intuitionistic Fuzzy Preference Relations." *Information Sciences*, *466*(October), 109–128. https://doi.org/10.1016/j.ins.2018.07.031.

Wang, Cheng-Yi, & Shyi-Ming Chen (2017). "Multiple Attribute Decision Making Based on Interval-Valued Intuitionistic Fuzzy Sets, Linear Programming Methodology, and the

Extended TOPSIS Method." *Information Sciences, 397–398*(August), 155–167. https://doi.org/10.1016/j.ins.2017.02.045.

Wang, Weize, & Xinwang Liu (2013). "An Extended LINMAP Method for Multi-Attribute Group Decision Making under Interval-Valued Intuitionistic Fuzzy Environment." *Procedia Computer Science*, First International Conference on Information Technology and Quantitative Management, *17*(January), 490–497. https://doi.org/10.1016/j.procs.2013.05.063.

Wei, Guiwu (2010). "Some Induced Geometric Aggregation Operators with Intuitionistic Fuzzy Information and Their Application to Group Decision Making." *Applied Soft Computing, 10*(2), 423–431. https://doi.org/10.1016/j.asoc.2009.08.009.

Wei, Gui-Wu (2010). "Extension of TOPSIS Method for 2-Tuple Linguistic Multiple Attribute Group Decision Making with Incomplete Weight Information." *Knowledge and Information Systems, 25*(3), 623–634. https://doi.org/10.1007/s10115-009-0258-3.

Won, Kwangjai, Eun-Sung Chung, & Sung-Uk Choi (2015). "Parametric Assessment of Water Use Vulnerability Variations Using SWAT and Fuzzy TOPSIS Coupled with Entropy." *Sustainability, 7*(9), 12052–12070. https://doi.org/10.3390/su70912052.

Wood, David A. (2016). "Supplier Selection for Development of Petroleum Industry Facilities, Applying Multi-Criteria Decision Making Techniques Including Fuzzy and Intuitionistic Fuzzy TOPSIS with Flexible Entropy Weighting." *Journal of Natural Gas Science and Engineering, 28*(January), 594–612. https://doi.org/10.1016/j.jngse.2015.12.021.

Wu, Jian, & Francisco Chiclana (2014). "Multiplicative Consistency of Intuitionistic Reciprocal Preference Relations and Its Application to Missing Values Estimation and Consensus Building." *Knowledge-Based Systems, 71*(November), 187–200. https://doi.org/10.1016/j.knosys.2014.07.024.

Wu, Jian, Yujia Liu, & Changyong Liang (2015). "A Consensus- and Harmony-Based Feedback Mechanism for Multiple Attribute Group Decision Making with Correlated Intuitionistic Fuzzy Sets." *International Transactions in Operational Research 22*(6), 1033–1054. https://doi.org/10.1111/itor.12143.

Xin, Jing (2010). "Optimization Model of Nuclear Emergency Decision-Making Proposal Based on Fuzzy Entropy Weight and TOPSIS." In *2010 Seventh International Conference on Fuzzy Systems and Knowledge Discovery*, 904–907. Yantai, China: IEEE. https://doi.org/10.1109/FSKD.2010.5569114.

Xu, Gai-li, Shu-ping Wan, Feng Wang, Jiu-ying Dong, & Yi-feng Zeng (2016). "Mathematical Programming Methods for Consistency and Consensus in Group Decision Making with Intuitionistic Fuzzy Preference Relations." *Knowledge-Based Systems, 98*(April), 30–43. https://doi.org/10.1016/j.knosys.2015.12.007.

Xu, Z. (2007). "Intuitionistic Fuzzy Aggregation Operators." *IEEE Transactions on Fuzzy Systems, 15*(6), 1179–1187. https://doi.org/10.1109/TFUZZ.2006.890678.

Xu, Z. S., & J. Chen (2008). "An Overview of Distance and Similarity Measures of Intuitionistic Fuzzy Sets." *International Journal of Uncertainty, Fuzziness and Knowledge-Based Systems, 16*(04), 529–555. https://doi.org/10.1142/S0218488508005406.

Xu, Zeshui (2012). *Intuitionistic Fuzzy Aggregation and Clustering*. Berlin, Heidelberg: Springer-Verlag. http://www.springer.com/us/book/9783642284052.

Xu, Zeshui, & Xiaoqiang Cai (2012). *Intuitionistic Fuzzy Information Aggregation: Theory and Applications*. Berlin, Heidelberg: Springer-Verlag. //www.springer.com/la/book/9783642295843.

Xu, Zeshui, & Ronald R. Yager (2006). "Some Geometric Aggregation Operators Based on Intuitionistic Fuzzy Sets." *International Journal of General Systems, 35*(4), 417–433. https://doi.org/10.1080/03081070600574353.

Xu, Zeshui, & Xiaolu Zhang (2013). "Hesitant Fuzzy Multi-Attribute Decision Making

Based on TOPSIS with Incomplete Weight Information." *Knowledge-Based Systems*, 52(November), 53–64. https://doi.org/10.1016/j.knosys.2013.05.011.

Yager, Ronald R. (2013). "Pythagorean Fuzzy Subsets." In *2013 Joint IFSA World Congress and NAFIPS Annual Meeting (IFSA/NAFIPS)*, 57–61. https://doi.org/10.1109/IFSA-NAFIPS.2013.6608375.

Yan, Du, & Zuo Jia (2011). "An Extended TOPSIS Method for the Multiple Attribute Group Decision Making Problems Based on Intuitionistic Linguistic Numbers." *Scientific Research and Essays*, 6(19), 4125–4132. https://doi.org/10.5897/SRE11.849.

Yang, Yong, & Xindong Peng (2017). "A Revised TOPSIS Method Based on Interval Fuzzy Soft Set Models with Incomplete Weight Information." *Fundamenta Informaticae*, 152(3), 297–321. https://doi.org/10.3233/FI-2017-1522.

Ye, Fei (2010). "An Extended TOPSIS Method with Interval-Valued Intuitionistic Fuzzy Numbers for Virtual Enterprise Partner Selection." *Expert Systems with Applications*, 37(10), 7050–7055. https://doi.org/10.1016/j.eswa.2010.03.013.

Ye, Jun (2013). "Multiple Attribute Group Decision-Making Methods with Completely Unknown Weights in Intuitionistic Fuzzy Setting and Interval-Valued Intuitionistic Fuzzy Setting." *Group Decision and Negotiation*, 22(2), 173–188. https://doi.org/10.1007/s10726-011-9255-5.

Ye, Jun (2015). "An Extended TOPSIS Method for Multiple Attribute Group Decision Making Based on Single Valued Neutrosophic Linguistic Numbers." *Journal of Intelligent & Fuzzy Systems*, 28(1), 247–255. https://doi.org/10.3233/IFS-141295.

Yuan, Yuan, & Li Yang He (2012). "Extension of TOPSIS for Multiple Attribute Decision Making Using Intuitionistic Fuzzy Sets." *Advanced Materials Research*, 433–440(January), 4053–4058. https://doi.org/10.4028/www.scientific.net/AMR.433-440.4053.

Zamri, Nurnadiah, & Lazim Abdullah (2014). "Flood Control Project Selection Using an Interval Type-2 Entropy Weight with Interval Type-2 Fuzzy TOPSIS." *AIP Conference Proceedings*, 1602(1), 62–68. https://doi.org/10.1063/1.4882467.

Zhang, Cheng, Huchang Liao, Li Luo, & Zeshui Xu (2020). "Distance-Based Consensus Reaching Process for Group Decision Making with Intuitionistic Multiplicative Preference Relations." *Applied Soft Computing*, 88(March), 106045. https://doi.org/10.1016/j.asoc.2019.106045.

Zhang, Fangwei, & Shihe Xu (2016). "Multiple Attribute Group Decision Making Method Based on Utility Theory Under Interval-Valued Intuitionistic Fuzzy Environment." *Group Decision and Negotiation*, 25(6), 1261–1275. https://doi.org/10.1007/s10726-016-9473-y.

Zhang, Hong, Chao-lin Gu, Lu-wen Gu, & Yan Zhang (2011). "The Evaluation of Tourism Destination Competitiveness by TOPSIS & Information Entropy – A Case in the Yangtze River Delta of China." *Tourism Management*, 32(2), 443–451. https://doi.org/10.1016/j.tourman.2010.02.007.

Zhang, Huimin, & Liying Yu (2012). "MADM Method Based on Cross-Entropy and Extended TOPSIS with Interval-Valued Intuitionistic Fuzzy Sets." *Knowledge-Based Systems*, 30(June), 115–120. https://doi.org/10.1016/j.knosys.2012.01.003.

Zhang, Liyuan, Tao Li, & Xuanhua Xu (2014). "Consensus Model for Multiple Criteria Group Decision Making under Intuitionistic Fuzzy Environment." *Knowledge-Based Systems*, 57(February), 127–135. https://doi.org/10.1016/j.knosys.2013.12.013.

Zhang, Xiaolu, & Zeshui Xu (2015). "Soft Computing Based on Maximizing Consensus and Fuzzy TOPSIS Approach to Interval-Valued Intuitionistic Fuzzy Group Decision Making." *Applied Soft Computing*, 26(January), 42–56. https://doi.org/10.1016/j.asoc.2014.08.073.

Zhang, Zhiming, & Witold Pedrycz (2018). "Goal Programming Approaches to Managing Consistency and Consensus for Intuitionistic Multiplicative Preference Relations in

Group Decision Making." *IEEE Transactions on Fuzzy Systems*, 26(6), 3261–3275. https://doi.org/10.1109/TFUZZ.2018.2818074.

Zhang, Zhiming, & Witold Pedrycz (2019). "A Consistency and Consensus-Based Goal Programming Method for Group Decision-Making With Interval-Valued Intuitionistic Multiplicative Preference Relations." *IEEE Transactions on Cybernetics*, 49(10), 3640–3654. https://doi.org/10.1109/TCYB.2018.2842073.

Zhao, Hua, Zeshui Xu, Mingfang Ni, & Shousheng Liu (2010). "Generalized Aggregation Operators for Intuitionistic Fuzzy Sets." *International Journal of Intelligent Systems*, 25(1), 1–30. https://doi.org/10.1002/int.20386.

8 Other Fuzzy TOPSIS Approaches

PYTHAGOREAN FUZZY SETS

Some authors consider intuitionistic fuzzy sets to be too restrictive, giving rise to other extensions such as Pythagorean fuzzy sets (Yager, 2013). Let X be a universe of discourse. A Pythagorean fuzzy set P in X is described by

$$P = \{<x, \mu_P(x), \vartheta_P(x) > | x \in X\}. \tag{8.1}$$

where $0 \leq \mu_P(x) \leq 1$ and $0 \leq \vartheta_P(x) \leq 1$.

However, unlike with intuitionistic fuzzy sets, $0 \leq (\mu_P(x))^2 + (\vartheta_P(x))^2 \leq 1$, and the indeterminacy degree becomes $\pi_P(x) = \sqrt{1 - (\mu_P(x))^2 - (\vartheta_P(x))^2}$.

A comparison between Pythagorean and intuitionistic fuzzy numbers spaces can be represented in Figure 8.1.

Hence, an intuitionistic fuzzy number is a Pythagorean fuzzy number, but not all Pythagorean fuzzy numbers are intuitionistic fuzzy numbers.

The distance between two Pythagorean fuzzy numbers $\alpha_1(\mu_1, \vartheta_1)$ and $\alpha_2(\mu_2, \vartheta_2)$ can be computed by (Biswas & Sarkar 2019)

$$d(\alpha_1, \alpha_2) = (0.5 \times (|\mu_1^2 - \mu_2^2|^\lambda + |\vartheta_1^2 - \vartheta_2^2|^\lambda + |\pi_1^2 - \pi_2^2|^\lambda))^{1/\lambda}, \tag{8.2}$$

where $\lambda = 1$ for the Hamming distance, $\lambda = 2$ for the Euclidean distance, and $\lambda = +\infty$ for the Chebyshev distance.

The distance satisfies

- $0 \leq d(\alpha_1, \alpha_2) \leq 1$
- $d(\alpha_1, \alpha_2) = 0 \leftrightarrow \alpha_1 = \alpha_2$
- $d(\alpha_1, \alpha_2) = d(\alpha_2, \alpha_1)$.

Let $\alpha_i = (\mu_i, \vartheta_i)$, with $1 \leq i \leq n$, be a collection of Pythagorean fuzzy numbers and ω_i be their relative weights. The Pythagorean Fuzzy Weighted Averaging operator is defined from $[0, 1]^{2n}$ to $[0, 1]^2$ as follows (Yager, 2014):

$$PFWA_\omega(\alpha_1, \alpha_2, \ldots, \alpha_n) = (\sum_{i=1}^{n} \omega_i \mu_i, \sum_{i=1}^{n} \omega_i \vartheta_i). \tag{8.3}$$

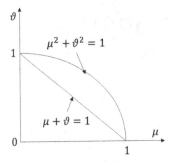

FIGURE 8.1 Comparison of Intuitionistic and Pythagorean Fuzzy-Number Spaces.

A Simple Consensus

While Pythagorean fuzzy sets can extend intuitionistic fuzzy sets, the relative computations are more complex (Yager, 2014). Hence the algorithm proposed by Hsu and Chen (1996) and extended by El Alaoui, Ben-azza, and Zahi (2016) is adopted here for achieving consensus.

Algorithm 8.1:

Step 1: Each decision maker k ($1 \leq k \leq K$) expresses their assessment using a Pythagorean fuzzy number α_k.

Step 2: Compute the distance between each pair of decision makers (Equation 8.2).

Step 3: Construct the agreement matrix $AM = [1 - d(\alpha_{k1}, \alpha_{k2})]_{K \times K}$.

Step 4: Compute the average agreement

$$AA_k = \frac{1}{K-1} \sum_{\substack{o=1 \\ o \neq k}}^{K} (1 - d(\alpha_k, \alpha_o)).$$

Step 5: Calculate the relative agreement

$$RA_k = \frac{AA_k}{\sum_{o=1}^{K} AA_o}.$$

Step 6: Compute the collective decision using the Pythagorean fuzzy weighted averaging operator (Equation 8.3):

$$\alpha = \left(\sum_{k=1}^{K} RA_k \mu_k, \sum_{k=1}^{K} RA_k \vartheta_i \right).$$

Pythagorean Fuzzy TOPSIS

A first attempt to adapt the TOPSIS algorithm using Pythagorean fuzzy sets was proposed by Zhang and Xu (2014). The problem treated contains only one decision maker, and the authors opted for a modified closeness coefficient taking values in the interval [−1, 1]. Yang et al. (2016) pointed out some errors and proposed corrections. The multiplicity of decision makers has been taken into consideration (Akram, Dudek, & Ilyas, 2019) with the same revised closeness coefficient as Zhang and Xu (2014).

Li et al. (2019) proposed three fuzzy Pythagorean similarities, investigated their properties, and used them to compute the closeness coefficients in order to present an extension of Pythagorean fuzzy TOPSIS. Further similarity-based methods in this context were proposed by Hussian and Yang (2019) and Rani et al. (2020).

Khan et al. (2020) proposed an extension using Pythagorean cubic fuzzy sets. They supposed a situation with incomplete weight information and used maximum deviation to find the optimal criterion weights.

Another approach to determining attribute weights was proposed by Han et al. (2019), based on entropy. The authors defined a fuzzy soft entropy that they judge more fit. The measure introduced integrates objective and subjective weights. Other entropy-based method for TOPSIS in the Pythagorean fuzzy context can be found (Biswas & Sarkar 2019; Han et al. 2020; Lin, Huang, & Xu 2019).

In an approach based on mathematical logic and set theory, Naeem, Riaz, and Afzal (2019) tried linking Pythagorean and m-polar fuzzy sets, an extension of bipolar fuzzy sets (Chen et al., 2014), to the TOPSIS algorithm. They argued that m-polar fuzzy sets do not take into consideration the nonmembership function, while Pythagorean fuzzy sets neglect polarity, which justifies merging the two (Hashmi & Riaz, 2020; Riaz & Hashmi, 2019).

Another combination linking rough sets (Zhang, Xie, & Wang, 2016) to Pythagorean fuzzy TOPSIS in a three-way decision approach (Yao, 2012) was presented by Liang et al. (2018), who used a loss function to evaluate the risk associated with each decision based on a Pythagorean fuzzy number, discussing the two situations when weight information is and is not known, and used maximum deviation in the latter.

Oz et al. (2019) proposed a Pythagorean fuzzy TOPSIS to prioritize risks according to occurrence and severity. The model proposed was applied to a natural-gas pipeline project.

Further approaches in the Pythagorean fuzzy context combine TOPSIS with AHP (Ak & Gul, 2019; Yildiz et al., 2020; Yucesan & Gul, 2020), COPRAS (Dorfeshan & Mousavi, 2019), and VIKOR (Naeem et al., 2019).

Similar to the extension of intuitionistic fuzzy sets to interval-valued intuitionistic fuzzy sets, Pythagorean fuzzy sets have been extended to interval-valued Pythagorean fuzzy sets with TOPSIS applications (Ho, Lin, & Chen 2019; Sajjad Ali Khan et al. 2018; Yu et al. 2019).

Proposed Algorithm

Algorithm 8.2:

Step 1: Each decision maker assesses each alternative according to qualitative criteria using a Pythagorean fuzzy number α_{ijk}.

Step 2: Each decision maker assesses each criterion's importance using a Pythagorean fuzzy number w_{jk}.

Step 3: Compute the group assessments for criterion weights w_j and alternatives α_{ij} using Algorithm 8.1.

Step 4: Construct the normalized decision matrix $R = [r_{ij}]_{m \times n}$.

For benefit criteria, $r_{ij} = \alpha_{ij} = (\mu_{ij}, \vartheta_{ij})$; for cost criteria, $r_{ij} = \alpha_{ij}^C = (\vartheta_{ij}, \mu_{ij})$.

Step 5: Construct the weighted normalized decision matrix (Equation 8.3).

Step 6: Compute the distances to the ideal solutions.

The distance to the positive ideal solution is

$$D_i^* = \sum_{j=1}^n 0.5 \times \left((r_{\mu_{ij}} - 1)^2 + r_{\vartheta_{ij}}^2\right), \quad i = 1, \ldots, m. \tag{8.4}$$

The distance to the negative ideal solution is

$$D_i^- = \sum_{j=1}^n 0.5 \times \left(r_{\mu_{ij}}^2 + (r_{\vartheta_{ij}} - 1)^2\right), \quad i = 1, \ldots, m. \tag{8.5}$$

Step 7: Compute the closeness coefficients

$$CC_i = \frac{D_i^-}{D_i^- + D_i^*}, \quad i = 1, \ldots, n. \tag{8.6}$$

Step 8: Rank the alternatives according to closeness coefficients, from highest to lowest.

Example

In an example treated by Biswas and Sarkar (2019) and Zhang (2016), five tables (alternatives A_1, A_2, A_3, A_4, and A_5) are evaluated by three decision makers (DM_1, DM_2, and DM_3) according to five criteria (C_1: features; C_2: display; C_3: communication; C_4: price; C_5: customer care).

Step 1: Each decision maker assesses each alternative according to qualitative criteria using a Pythagorean fuzzy number (Table 8.1).

TABLE 8.1
Individual Fuzzy Pythagorean Assessments

Decision Maker	Alternative	Criterion				
		C_1	C_2	C_3	C_4	C_5
DM_1	A_1	(0.8, 0.4)	(0.8, 0.6)	(0.6, 0.7)	(0.8, 0.3)	(0.6, 0.5)
	A_2	(0.5, 0.7)	(0.9, 0.2)	(0.8, 0.5)	(0.6, 0.3)	(0.5, 0.6)
	A_3	(0.4, 0.3)	(0.3, 0.7)	(0.7, 0.4)	(0.4, 0.6)	(0.5, 0.4)
	A_4	(0.6, 0.6)	(0.7, 0.5)	(0.7, 0.2)	(0.6, 0.4)	(0.7, 0.3)
	A_5	(0.7, 0.5)	(0.6, 0.4)	(0.9, 0.3)	(0.7, 0.6)	(0.7, 0.1)
DM_2	A_1	(0.9, 0.3)	(0.7, 0.6)	(0.5, 0.8)	(0.6, 0.3)	(0.6, 0.3)
	A_2	(0.4, 0.7)	(0.9, 0.2)	(0.8, 0.1)	(0.5, 0.3)	(0.5, 0.3)
	A_3	(0.6, 0.3)	(0.7, 0.7)	(0.7, 0.6)	(0.4, 0.4)	(0.3, 0.4)
	A_4	(0.8, 0.4)	(0.7, 0.5)	(0.6, 0.2)	(0.7, 0.4)	(0.7, 0.4)
	A_5	(0.7, 0.2)	(0.8, 0.2)	(0.8, 0.4)	(0.6, 0.6)	(0.6, 0.6)
DM_3	A_1	(0.8, 0.6)	(0.7, 0.6)	(0.5, 0.8)	(0.5, 0.5)	(0.6, 0.1)
	A_2	(0.5, 0.6)	(0.9, 0.2)	(0.8, 0.1)	(0.5, 0.3)	(0.4, 0.3)
	A_3	(0.7, 0.4)	(0.7, 0.5)	(0.6, 0.1)	(0.9, 0.2)	(0.5, 0.6)
	A_4	(0.9, 0.2)	(0.5, 0.6)	(0.6, 0.2)	(0.6, 0.1)	(0.7, 0.4)
	A_5	(0.6, 0.1)	(0.8, 0.2)	(0.9, 0.2)	(0.5, 0.6)	(0.6, 0.4)

Step 2: Each decision maker assesses each criterion's importance using a Pythagorean fuzzy number.

To ease computation and enable comparison, criterion weights are considered here (as in Zhang 2016) to be given by crisp numbers: (0.2, 0.4, 0.1, 0.1, 0.2).

Step 3: Compute the group assessments for criterion weights and alternatives using Algorithm 8.1 (Table 8.2).

Step 4: Construct the normalized decision matrix (Table 8.3).

Step 5: Construct the weighted normalized decision matrix (Table 8.4).

Step 6: Compute the distances to the ideal solutions (Table 8.5).

Step 7: Compute the closeness coefficients (Table 8.5).

Step 8: Rank the alternatives according to closeness coefficients, from highest to lowest (Table 8.5).

The proposed ranking and those proposed by Biswas and Sarkar (2019) and Zhang (2016) are similar—one permutation away, between A_3 and A_4 ranked third and fourth. The difference is due to several causes: the choice of fixed or relative ideal solutions, the distances used, and the last-aggregation approach used by those authors. A comparison between first and last aggregation for TOPSIS was proposed by Roghanian, Rahimi, and Ansari (2010).

NEUTROSOPHIC SETS

Neutrosophic sets (Smarandache, 1999) are also a generalization of intuitionistic fuzzy sets that permit dealing with inconsistent information. Let X be a universe of

TABLE 8.2
Group Fuzzy Pythagorean Assessments

Alternative	Criterion				
	C_1	C_2	C_3	C_4	C_5
A_1	(0.8331, 0.4318)	(0.7315, 0.6)	(0.5315, 0.7685)	(0.6279, 0.3668)	(0.6, 0.2968)
A_2	(0.4664, 0.6672)	(0.9, 0.2)	(0.8, 0.2206)	(0.5320, 0.3)	(0.4660, 0.3901)
A_3	(0.5682, 0.3318)	(0.5776, 0.6306)	(0.6696, 0.3735)	(0.5167, 0.43)	(0.4351, 0.4632)
A_4	(0.7724, 0.3948)	(0.6397, 0.5302)	(0.6318, 0.2)	(0.6326, 0.3035)	(0.7, 0.3675)
A_5	(0.6675, 0.2617)	(0.7410, 0.2590)	(0.8682, 0.2977)	(0.5996, 0.6)	(0.6326, 0.3652)

TABLE 8.3
Normalized Group Fuzzy Pythagorean Assessments

Alternative	Criterion				
	C_1	C_2	C_3	C_4	C_5
A_1	(0.4318, 0.8331)	(0.7315, 0.6)	(0.5315, 0.7685)	(0.3668, 0.6279)	(0.2968, 0.6)
A_2	(0.6672, 0.4664)	(0.9, 0.2)	(0.8, 0.2206)	(0.3, 0.5320)	(0.3901, 0.4660)
A_3	(0.3318, 0.5682)	(0.5776, 0.6306)	(0.6696, 0.3735)	(0.43, 0.5167)	(0.4632, 0.4351)
A_4	(0.3948, 0.7724)	(0.6397, 0.5302)	(0.6318, 0.2)	(0.3035, 0.6326)	(0.3675, 0.7)
A_5	(0.2617, 0.6675)	(0.7410, 0.2590)	(0.8682, 0.2977)	(0.6, 0.5996)	(0.3652, 0.6326)

discourse. A neutrosophic set N in X is described by the truth function $T_N(x)$, the indeterminacy function $I_N(x)$, and the falsity function $F_N(x)$ as follows:

$$N = \{<x, T_N(x), I_N(x), F_N(x) > | x \in X\}, \tag{8.7}$$

where $0 \leq T_N(x) \leq 1$, $0 \leq I_N(x) \leq 1$, and $0 \leq F_N(x) \leq 1$.

TABLE 8.4
Weighted Normalized Group Fuzzy Pythagorean Assessments

Alternative	Criterion				
	C_1	C_2	C_3	C_4	C_5
A_1	(0.0846, 0.1666)	(0.2926, 0.2400)	(0.0531, 0.0769)	(0.0367, 0.0628)	(0.0594, 0.1200)
A_2	(0.1334, 0.0933)	(0.3600, 0.0800)	(0.0800, 0.0221)	(0.0300, 0.0532)	(0.0780, 0.0932)
A_3	(0.0664, 0.1136)	(0.2310, 0.2522)	(0.0670, 0.0374)	(0.0430, 0.0517)	(0.0926, 0.0870)
A_4	(0.0790, 0.1545)	(0.2559, 0.2121)	(0.0632, 0.0200)	(0.0303, 0.0633)	(0.0735, 0.1400)
A_5	(0.0523, 0.1335)	(0.2964, 0.1036)	(0.0868, 0.0298)	(0.0600, 0.0600)	(0.0730, 0.1265)

TABLE 8.5
Distances to Ideal Solutions, Closeness Coefficients, and Final and Compared Rankings

Alternative	D_i^-	D_i^*	CC_i	Proposed Rank	Rank from Biswas and Sarkar (2019) and Zhang (2016)
A_1	2.0770	1.9389	0.4828	5	5
A_2	1.9125	2.2522	0.5408	1	1
A_3	2.0805	2.0385	0.4949	3	4
A_4	2.0856	1.9976	0.4892	4	3
A_5	2.0095	2.1247	0.5139	2	2

However, unlike intuitionistic and Pythagorean fuzzy sets, the three functions are completely independent of each other: $0 \leq T_N(x) + I_N(x) + F_N(x) \leq 3$.

Similarly to interval-valued intuitionistic and Pythagorean fuzzy sets, neutrosophic sets have been extended into interval neutrosophic sets (Wang et al., 2005). Hence, the truth function, indeterminacy function, and falsity function are represented by intervals satisfying

$$T_N(x) = [T_N^L(x), T_N^U(x)] \subseteq [0, 1], \ I_N(x) = [I_N^L(x), I_N^U(x)] \subseteq [0, 1] \text{ and } F_N(x) = [F_N^L(x) F_N^U(x)] \subseteq [0, 1].$$

For convenience, a neutrosophic number will be noted $N([T_N^L, T_N^U], [I_N^L, I_N^U], [F_N^L, F_N^U])$ (Peng, 2019).

Let $N_1([T_1^L, T_1^U], [I_1^L, I_1^U], [F_1^L, F_1^U])$ and $N_2([T_2^L, T_2^U], [I_2^L, I_2^U], [F_2^L, F_2^U])$ be two interval neutrosophic numbers; their Hamming distance is defined as Peng (2019)

$$d_{HN}(N_1, N_2) = (1/6) \times (|T_1^L - T_2^L| + |T_1^U - T_2^U| + |I_1^L - I_2^L| + |I_1^U - I_2^U| + |F_1^L - F_2^L| + |F_1^U - F_2^U|). \tag{8.8}$$

Let $N_i([T_i^L, T_i^U], [I_i^L, I_i^U], [F_i^L, F_i^U])$, with $1 \leq i \leq n$, be a collection of interval neutrosophic numbers and ω_i be the relative weight of each N_i. The Interval Neutrosophic Weighted Averaging (INWA) operator is defined from $[0, 1]^{6n}$ to $[0, 1]^6$ as follows:

$$INWA_\omega(N_1, \ldots, N_n) = ([1 - \prod_{i=1}^n (1 - T_i^L)^{\omega_i}, 1 - \prod_{i=1}^n (1 - T_i^U)^{\omega_i}], \\ [\prod_{i=1}^n (I_i^L)^{\omega_i}, \prod_{i=1}^n (I_i^U)^{\omega_i}], [\prod_{i=1}^n (F_i^L)^{\omega_i}, \prod_{i=1}^n (F_i^U)^{\omega_i}]). \tag{8.9}$$

As pointed out in the previous chapter for intuitionistic and interval-valued intuitionistic fuzzy sets (El Alaoui & Ben-azza, 2017), a single $T_i^L = 1$ will force the aggregated truth function to 1 regardless of the other aggregated entries. Similar veto behavior can be seen with $T_i^U = 1$, $I_i^L = 0$, I_i^U, $F_i^L = 0$, or $F_i^U = 0$. Thus, the Generalized INWA (GINWA) can be computed as

$$GINWA_\omega(N_1, \ldots, N_n) = ([AT^L, AT^U], [AI^L, AI^U], [AF^L, AF^U]), \tag{8.10}$$

where

$$AT^L = 1 - \frac{\prod_{i=1}^n (2 - w_i - T_i^L)^{1/n} - \prod_{i=1}^n (1 - w_i)^{1/n}}{\prod_{i=1}^n (2 - w_i)^{1/n} - \prod_{i=1}^n (1 - w_i)^{1/n}}$$

$$AT^U = 1 - \frac{\prod_{i=1}^n (2 - w_i - T_i^U)^{1/n} - \prod_{i=1}^n (1 - w_i)^{1/n}}{\prod_{i=1}^n (2 - w_i)^{1/n} - \prod_{i=1}^n (1 - w_i)^{1/n}}$$

$$AI^L = \frac{\prod_{i=1}^n (1 - w_i + I_i^L)^{1/n} - \prod_{i=1}^n (1 - w_i)^{1/n}}{\prod_{i=1}^n (2 - w_i)^{1/n} - \prod_{i=1}^n (1 - w_i)^{1/n}}$$

$$AI^U = \frac{\prod_{i=1}^n (1 - w_i + I_i^U)^{1/n} - \prod_{i=1}^n (1 - w_i)^{1/n}}{\prod_{i=1}^n (2 - w_i)^{1/n} - \prod_{i=1}^n (1 - w_i)^{1/n}}$$

$$AF^L = \frac{\prod_{i=1}^n (1 - w_i + F_i^L)^{1/n} - \prod_{i=1}^n (1 - w_i)^{1/n}}{\prod_{i=1}^n (2 - w_i)^{1/n} - \prod_{i=1}^n (1 - w_i)^{1/n}}$$

$$AF^U = \frac{\prod_{i=1}^n (1 - w_i + F_i^U)^{1/n} - \prod_{i=1}^n (1 - w_i)^{1/n}}{\prod_{i=1}^n (2 - w_i)^{1/n} - \prod_{i=1}^n (1 - w_i)^{1/n}}$$

Hence, in this context Algorithm 8.1 becomes the following:

Algorithm 8.3:

Step 1: Each decision maker k ($1 \leq k \leq K$) expresses their assessment using an interval neutrosophic number $N_k([T_k^L, T_k^U], [I_k^L, I_k^U], [F_k^L, F_k^U])$.

Step 2: Compute the distance between each pair of decision makers (Equation 8.8).

Step 3: Construct the agreement matrix $AM = [1 - d(\alpha_{k1}, \alpha_{k2})]_{K \times K}$.

Step 4: Compute the average agreement

$$AA_k = \frac{1}{K-1} \sum_{\substack{o=1 \\ o \neq k}}^{K} (1 - d(\alpha_k, \alpha_o)).$$

Step 5: Calculate the relative agreement

$$RA_k = \frac{AA_k}{\sum_{o=1}^{K} AA_o}.$$

Step 6: Compute the collective decision using GINWA (Equation 8.10).

Reviews of neutrosophic sets have been proposed by El-Hefenawy et al. (2016), Khan et al. (2018), Otay and Kahraman (2019), and Peng and Dai (2020).

Neutrosophic TOPSIS

Several attempts to adapt TOPSIS to the neutrosophic context exist in the literature, with applications including investment risk evaluation (Liang, Zhao, & Wu, 2017), green-supplier evaluation (Chen, Zeng, & Zhang, 2018), smart medical-device selection (Abdel-Basset, Manogaran et al., 2019), unmanned aerial-vehicle evaluation (Karaşan & Kaya, 2020), and personnel selection (Nabeeh et al., 2019).

Tehrim and Riaz (2019) proposed a TOPSIS extension using bipolar neutrosophic soft sets. The proposed approach was topologically oriented, discussing the space closure, interior, exterior, and frontier.

Supposing that attribute weights are partially known or completely unknown, Selvachandran et al. (2018) proposed a maximum-deviation approach to deduce their values, considering only single-valued neutrosophic sets. Other TOPSIS approaches using single-valued neutrosophic sets have been given by Biswas, Pramanik, and Giri (2016), Nancy and Garg (2019), and Zeng et al. (2020).

Further TOPSIS approaches used interval neutrosophic sets (Otay & Kahraman, 2018; Sharma et al., 2019) and type 2 neutrosophic sets (Abdel-Basset, Saleh et al., 2019; Karaaslan & Hunu, 2020).

In addition, some hybrid extensions using neutrosophic TOPSIS were proposed in combination with AHP (Junaid et al., 2020; Karaşan, Bolturk, & Kahraman, 2020), ANP (Abdel-Basset, Mohamed, & Smarandache, 2018), ELECTRE (Akram,

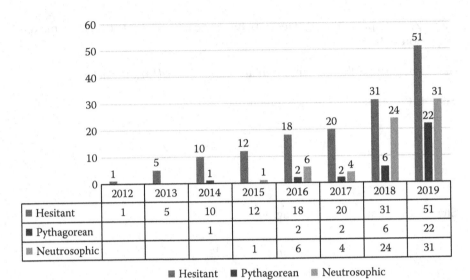

FIGURE 8.2 Neutrosophic, Hesitant, and Pythagorean Fuzzy TOPSIS in the Scopus Database, 2012–2019.

Shumaiza, and Smarandache, 2018), MABAC (Peng, & Dai, 2018), and DEMATEL (Kilic & Yalcin, 2020; Yang & Pang, 2018).

Figure 8.2 reviews publications on hesitant, Pythagorean, and neutrosophic TOPSIS (among other types reviewed in this chapter) in the Scopus database through 2019.

Proposed Algorithm

Algorithm 8.4:

Step 1: Each decision maker assesses each alternative according to qualitative criteria using linguistic variables that will be converted to interval neutrosophic number N_{ijk}.

Step 2: Each decision maker assesses each criterion's importance using linguistic variables that will be converted to interval neutrosophic number w_{jk}.

Step 3: Compute the group assessments for criterion weights w_j and alternatives N_{ij} using Algorithm 8.3.

Step 4: Construct the weighted normalized decision matrix $V_{ij}([T_{V_{ij}}^L, T_{V_{ij}}^U], [I_{V_{ij}}^L, I_{V_{ij}}^U], [F_{V_{ij}}^L, F_{V_{ij}}^U])$ with

$$T_{V_{ij}}^L = T_{N_{ij}}^L, T_{V_{ij}}^U = T_{N_{ij}}^U, I_{V_{ij}}^L = I_{N_{ij}}^L, I_{V_{ij}}^U = I_{N_{ij}}^U, F_{V_{ij}}^L = F_{N_{ij}}^L \text{ and } F_{V_{ij}}^U = F_{N_{ij}}^U. \quad (8.11)$$

Step 5: Determine the ideal solutions.

Other Fuzzy TOPSIS Approaches

The positive ideal solution is

$$A^* = \{v_1^*, \ldots, v_m^*\}, \tag{8.12}$$

with $\tilde{v}_j^* = ([1, 1], [0, 0], [0, 0])$ for benefit criteria and $\tilde{v}_j^* = ([0, 0], [0, 0], [1, 1])$ for cost criteria.

The negative ideal solution is

$$A^* = \{v_1^-, \ldots, v_m^-\}, \tag{8.13}$$

with $\tilde{v}_j^- = ([0, 0], [0, 0], [1, 1])$ for benefit criteria and $\tilde{v}_j^- = ([1, 1], [0, 0], [0, 0])$ for cost criteria.

Step 6: Compute the distance between each alternative and the ideal solutions.

The distance to the positive ideal solution is

$$D_i^* = \sum_{j=1}^{m} d_{HN}(V_{ij}, \tilde{v}_j^*), \quad i = 1, \ldots, n, \tag{8.14}$$

while the distance to the negative ideal solution is

$$D_i^- = \sum_{j=1}^{m} d_{HN}(V_{ij}, \tilde{v}_j^-), \quad i = 1, \ldots, n. \tag{8.15}$$

Step 7: Compute the closeness coefficients

$$CC_i = \frac{D_i^-}{D_i^- + D_i^*}, \quad i = 1, \ldots, n. \tag{8.16}$$

Step 8: Rank the alternatives according to the closeness coefficients, from highest to lowest.

EXAMPLE

In this example treated by Biswas, Pramanik, and Giri (2019), four medical representatives (alternatives A_1, A_2, A_3, and A_4) are evaluated by four decision makers (DM_1, DM_2, DM_3, and DM_4) according to four criteria (C_1: oral communication skill; C_2: past experience; C_3: general aptitude; C_4: self-confidence).

Step 1: Each decision maker assesses each alternative using a linguistic variable N_{ijk} (Tables 8.6 and 8.7).

TABLE 8.6
Linguistic Variables for Assessing Alternatives and Criterion Importance

Linguistic Variable for Alternatives	Linguistic Variable for Criteria	Interval Neutrosophic Numbers
Extremely low (EL)	Extremely poor (EP)	([0, 0], [0, 0], [1, 1])
Very low (VL)	Very poor (VP)	([0.1, 0.15], [0.8, 0.85], [0.85, 0.9])
Low (L)	Poor (P)	([0.35, 0.4], [0.6, 0.65], [0.7, 0.75])
Medium (M)	Fair (F)	([0.5, 0.55], [0.4, 0.45], [0.45, 0.5])
High (H)	Good (G)	([0.75, 0.8], [0.15, 0.2], [0.1, 0.15])
Very high (VH)	Very good (VG)	([0.85, 0.9], [0.1, 0.15], [0.05, 0.1])
Extremely high (EH)	Extremely good (EG)	([1, 1], [0, 0], [0, 0])

TABLE 8.7
Linguistic Assessments of Four Alternatives According to Four Criteria by Four Decision Makers

Alternative	Decision Maker	Criterion			
		C_1	C_2	C_3	C_4
A_1	DM_1	VG	M	G	VG
	DM_2	G	M	G	G
	DM_3	G	G	VG	M
	DM_4	VG	M	G	G
A_2	DM_1	M	G	VG	G
	DM_2	G	M	VG	VG
	DM_3	M	G	G	G
	DM_4	G	G	VG	G
A_3	DM_1	M	M	G	VG
	DM_2	M	M	G	G
	DM_3	G	M	M	M
	DM_4	M	G	VG	M
A_4	DM_1	M	G	G	M
	DM_2	G	M	M	G
	DM_3	G	G	G	M
	DM_4	M	M	G	G

Step 2: Each decision maker assesses each criterion's importance using a linguistic variable w_{jk} (Table 8.8).
Step 3: Compute the group assessments for criterion weights w_j and alternatives N_{ij} using Algorithm 8.3 (Table 8.9).
Step 4: Construct the weighted normalized group decision matrix (Table 8.10).
Step 5: Determine the ideal solutions.

Since all criteria are benefit criteria, the positive ideal solution is {([1, 1], [0, 0], [0, 0]), ([1, 1], [0, 0], [0, 0]), ([1, 1], [0, 0], [0, 0]), ([1, 1], [0, 0], [0,

TABLE 8.8
Individual Assessments of Four Criterion Weights by Four Decision Makers

Decision Maker	Criterion			
	C_1	C_2	C_3	C_4
DM_1	VH	F	H	H
DM_2	H	F	H	H
DM_3	H	H	VH	F
DM_4	F	F	H	H

TABLE 8.9
Fuzzy Group Decision Matrix

Alternative	Criterion			
	C_1	C_2	C_3	C_4
A_1	([0.5939, 0.6442], [0.3205, 0.3708], [0.3402, 0.3906])	([0.6320, 0.6823], [0.2673, 0.3177], [0.2600, 0.3107])	([0.7758, 0.8258], [0.1374, 0.1874], [0.0873, 0.1374])	([0.7498, 0.8004], [0.1771, 0.2276], [0.1452, 0.1960])
A_2	([0.6947, 0.7450], [0.2046, 0.2550], [0.1374, 0.2240])	([0.5000, 0.5500], [0.4000, 0.4500], [0.4500, 0.5000])	([0.7222, 0.7726], [0.1911, 0.2415], [0.1595, 0.2102])	([0.7758, 0.8258], [0.1374, 0.1874], [0.0873, 0.1374])
A_3	([0.6947, 0.7450], [0.2046, 0.2550], [0.1374, 0.2240])	([0.6947, 0.7450], [0.2046, 0.2550], [0.1734, 0.2240])	([0.7222, 0.7726], [0.1911, 0.2415], [0.1595, 0.2102])	([0.5654, 0.6155], [0.3343, 0.3845], [0.3547, 0.4051])
A_4	([0.6603, 0.7108], [0.2535, 0.3039], [0.2456, 0.2964])	([0.6320, 0.6823], [0.2673, 0.3177], [0.2600, 0.3107])	([0.8013, 0.8514], [0.1246, 0.1747], [0.0746, 0.1246])	([0.6947, 0.7450], [0.2046, 0.2550], [0.1734, 0.2240])
Weights	([0.7222, 0.7726], [0.1911, 0.2415], [0.1595, 0.2102])	([0.5654, 0.6155], [0.3343, 0.3845], [0.3547, 0.4051])	([0.7758, 0.8258], [0.1374, 0.1874], [0.0873, 0.1374])	([0.6947, 0.7450], [0.2046, 0.2550], [0.1734, 0.2240])

TABLE 8.10
Weighted Normalized Group Decision Matrix

Alternative	Criterion			
	C_1	C_2	C_3	C_4
A_1	([0.4289, 0.4977], [0.0612, 0.0895], [0.0543, 0.0821])	([0.3573, 0.4200], [0.0894, 0.1221], [0.0922, 0.1258])	([0.6018, 0.6820], [0.0189, 0.0351], [0.0076, 0.0189])	([0.5209, 0.5963], [0.0362, 0.0580], [0.0252, 0.0439])
A_2	([0.5017, 0.5757], [0.0391, 0.0616], [0.0277, 0.0471])	([0.2827, 0.3385], [0.1337, 0.1730], [0.1596, 0.2025])	([0.5602, 0.6380], [0.0262, 0.0452], [0.0139, 0.0289])	([0.0589, 0.6153], [0.0281, 0.0478], [0.0151, 0.0308])
A_3	([0.5017, 0.5757], [0.0391, 0.0616], [0.0277, 0.0471])	([0.3928, 0.4586], [0.0684, 0.0980], [0.0615, 0.0907])	([0.5602, 0.6380], [0.0262, 0.0452], [0.0139, 0.0289])	([0.3928, 0.4586], [0.0684, 0.0980], [0.0615, 0.0907])
A_4	([0.4769, 0.5492], [0.0484, 0.0734], [0.0392, 0.0623])	([0.3573, 0.4200], [0.0894, 0.1221], [0.0922, 0.1258])	([0.6216, 0.7031], [0.0171, 0.0327], [0.0065, 0.0171])	([0.4826, 0.5551], [0.0419, 0.0650], [0.0301, 0.0502])

0], [0, 0])} and the negative ideal solution is {([0, 0], [0, 0], [1, 1]), ([0, 0], [0, 0], [1, 1]), ([0, 0], [0, 0], [1, 1]), ([0, 0], [0, 0], [1, 1])}.

Step 6: Compute the distance between each alternative and the ideal solutions (Table 8.11).

Step 7: Compute the closeness coefficients (Table 8.11).

Step 8: Rank the alternatives according to the closeness coefficients, from highest to lowest (Table 8.11).

In addition to the choice of ideal solutions (El Alaoui, El Yassini, & Ben-azza, 2019) and the aggregation operator used, the difference between the ranking proposed here and that of Biswas, Pramanik, and Giri (2019) can also be explained by the calculation of decision-maker weights according to each criteria, which can better represent reality.

HESITANT FUZZY SETS

Hesitant fuzzy sets are a different approach to fuzziness (Torra, 2010). Suppose a decision maker estimates that an alternative satisfies one criterion at a certain degree, say 0.3, a second criterion at 0.4, and a third criterion at 0.6. This situation can be described by the hesitant fuzzy number $H\{0.3, 0.4, 0.6\}$ (Xu, 2014; Zhang & Xu, 2017). Reviews of hesitant fuzzy sets have been given (Liao et al., 2018, 2020; Rodríguez et al., 2014, 2016), and further specialized books on the topic have been published (Farhadinia & Xu, 2019; Liao & Xu, 2017; Zhou & Xu, 2020).

Similar to other types of fuzzy sets, hesitant fuzzy sets have been extended to interval hesitant fuzzy sets (Chen, Xu, & Xia, 2013), in which the indeterminacy is with regard to not the value assigned but the membership assigned to the evaluation,

TABLE 8.11
Distances to Ideal Solutions, Closeness Coefficients, and Final Ranking

Alternative	D_i^-	D_i^*	CC_i	Rank
A_1	0.8093	2.0276	0.7147	2
A_2	0.8382	2.0134	0.7061	4
A_3	0.8248	2.0102	0.7091	3
A_4	0.7913	2.0387	0.7204	1

giving rise to a multitude of interval-valued hesitant TOPSIS methods (Akram & Adeel, 2019; Joshi & Kumar, 2018; Joshi & Kumar, 2016; Liu & Chen, 2017; Tavakkoli-Moghaddam et al., 2015). Hesitant fuzzy sets have also been extended to type 2 and interval type 2 hesitant fuzzy sets (Feng, Chuan-qiang, & Wei-he, 2018; Onar, Oztaysi, & Kahraman, 2014), with relevant TOPSIS applications (Chen et al., 2019).

Applications of hesitant fuzzy TOPSIS include hospital-site selection (Senvar, Otay, & Bolturk, 2016), lean performance assessment (Pérez-Domínguez et al., 2019), firm selection (Estrella et al., 2017), robot selection (Deli, 2020), deal selection (Amin, Fahmi, & Abdullah, 2019), smart glass evaluation (Büyüközkan & Güler, 2018), and more (Bhaumik, Roy, & Weber, 2020; Fahmi et al., 2019).

It should be noted that the duality between probability (De, Das, & Kar, 2019; Wu et al., 2019; Xian & Guo, 2020) and possibility (Wu et al., 2019; Wu, Chen, & Xu, 2017; Zhang, 2017) is all the more persistent in hesitant fuzzy TOPSIS methods.

Further TOPSIS approaches combining hesitant fuzzy sets with neutrosophic sets have been proposed (Akram, Naz, & Smarandache, 2019; Fahmi, Aslam, & Abdullah, 2019), as have combinations of fuzzy sets with Pythagorean sets (Liang & Xu 2017; Sajjad Ali Khan et al. 2018; Wang et al. 2019).

Hybrid hesitant methods combining TOPSIS and AHP have been proposed (Beskese et al., 2020; Onar et al., 2014), as have hybrids of TOPSIS and VIKOR (Wu et al. 2019). Figure 8.2 shows the prevalence of works addressing "TOPSIS" and "Hesitant" in the Scopus database through 2019.

Of course, other fuzzy theories and relevant adaptations in TOPSIS exist in the literature; but this book focuses on the most-used and emerging ones. Further fuzzy TOPSIS approaches may be included in future editions.

CONCLUSION

Due to the mathematical properties of intuitionistic fuzzy sets, several authors consider them too restrictive, giving rise to several extensions. This chapter reviews the extensions of intuitionistic fuzzy sets to Pythagorean fuzzy sets and neutrosophic sets. It investigates the classically used aggregation operators, proposes improved ones, and presents TOPSIS extensions including adapted consensus-reaching methods with relevant examples. The chapter also discusses TOPSIS extensions with hesitant fuzzy sets.

REFERENCES

Abdel-Basset, Mohamed, Gunasekaran Manogaran, Abduallah Gamal, & Florentin Smarandache (2019). "A Group Decision Making Framework Based on Neutrosophic TOPSIS Approach for Smart Medical Device Selection." *Journal of Medical Systems*, *43*(2), 38. https://doi.org/10.1007/s10916-019-1156-1.

Abdel-Basset, Mohamed, Mai Mohamed, & Florentin Smarandache (2018). "A Hybrid Neutrosophic Group ANP-TOPSIS Framework for Supplier Selection Problems." *Symmetry*, *10*(6), 226. https://doi.org/10.3390/sym10060226.

Abdel-Basset, Mohamed, M. Saleh, Abduallah Gamal, and Florentin Smarandache (2019). "An Approach of TOPSIS Technique for Developing Supplier Selection with Group Decision Making under Type-2 Neutrosophic Number." *Applied Soft Computing*, *77*(April), 438–452. https://doi.org/10.1016/j.asoc.2019.01.035.

Ak, M. Fatih, & Muhammet Gul (2019). "AHP–TOPSIS Integration Extended with Pythagorean Fuzzy Sets for Information Security Risk Analysis." *Complex & Intelligent Systems*, *5*(2), 113–126. https://doi.org/10.1007/s40747-018-0087-7.

Akram, Muhammad, & Arooj Adeel (2019). "TOPSIS Approach for MAGDM Based on Interval-Valued Hesitant Fuzzy N-Soft Environment." *International Journal of Fuzzy Systems*, *21*(3), 993–1009. https://doi.org/10.1007/s40815-018-0585-1.

Akram, Muhammad, Wieslaw A. Dudek, & Farwa Ilyas (2019). "Group Decision-Making Based on Pythagorean Fuzzy TOPSIS Method." *International Journal of Intelligent Systems*, *34*(7), 1455–1475. https://doi.org/10.1002/int.22103.

Akram, Muhammad, Sumera Naz, & Florentin Smarandache (2019). "Generalization of Maximizing Deviation and TOPSIS Method for MADM in Simplified Neutrosophic Hesitant Fuzzy Environment." *Symmetry*, *11*(8), 1058. https://doi.org/10.3390/sym11081058.

Akram, Muhammad, Shumaiza, & Florentin Smarandache (2018). "Decision-Making with Bipolar Neutrosophic TOPSIS and Bipolar Neutrosophic ELECTRE-I." *Axioms*, *7*(2), 33. https://doi.org/10.3390/axioms7020033.

Amin, Fazli, Aliya Fahmi, and Saleem Abdullah (2019). "Dealer Using a New Trapezoidal Cubic Hesitant Fuzzy TOPSIS Method and Application to Group Decision-Making Program." *Soft Computing*, *23*(14), 5353–5366. https://doi.org/10.1007/s00500-018-3476-3.

Beskese, Ahmet, Alper Camci, Gul Tekin Temur, & Ercan Erturk (2020). "Wind Turbine Evaluation Using the Hesitant Fuzzy AHP-TOPSIS Method with a Case in Turkey." *Journal of Intelligent & Fuzzy Systems*, *38*(1), 997–1011. https://doi.org/10.3233/JIFS-179464.

Bhaumik, Ankan, Sankar Kumar Roy, & Gerhard Wilhelm Weber (2020). "Hesitant Interval-Valued Intuitionistic Fuzzy-Linguistic Term Set Approach in Prisoners' Dilemma Game Theory Using TOPSIS: A Case Study on Human-Trafficking." *Central European Journal of Operations Research*, *28*(2), 797–816. https://doi.org/10.1007/s10100-019-00638-9.

Biswas, Animesh, & Biswajit Sarkar (2019). "Pythagorean Fuzzy TOPSIS for Multicriteria Group Decision-Making with Unknown Weight Information through Entropy Measure: BISWAS AND SARKAR." *International Journal of Intelligent Systems*, *34*(6), 1108–1128. https://doi.org/10.1002/int.22088.

Biswas, Pranab, Surapati Pramanik, & Bibhas C. Giri (2016). "TOPSIS Method for Multi-Attribute Group Decision-Making under Single-Valued Neutrosophic Environment." *Neural Computing and Applications*, *27*(3), 727–737. https://doi.org/10.1007/s00521-015-1891-2.

Biswas, Pranab, Surapati Pramanik, & Bibhas C. Giri (2019). "Neutrosophic TOPSIS with Group Decision Making." In *Fuzzy Multi-Criteria Decision-Making Using Neutrosophic Sets*, edited by Cengiz Kahraman and İrem Otay, (pp. 543–585). Studies in Fuzziness and Soft Computing. Cham: Springer International Publishing. https://doi.org/10.1007/978-3-030-00045-5_21.

Büyüközkan, Gülçin, & Merve Güler (2018). "A Hesitant Fuzzy Based TOPSIS Approach for Smart Glass Evaluation." In *Advances in Fuzzy Logic and Technology 2017*, edited byJanusz Kacprzyk, Eulalia Szmidt, Sławomir Zadrożny, Krassimir T. Atanassov, & Maciej Krawczak, (pp. 330–341). Advances in Intelligent Systems and Computing. Cham: Springer International Publishing. https://doi.org/10.1007/978-3-319-66830-7_30.

Chen, Juanjuan, Shenggang Li, Shengquan Ma, & Xueping Wang (2014). "M-Polar Fuzzy Sets: An Extension of Bipolar Fuzzy Sets." *The Scientific World Journal 2014*. https://doi.org/10.1155/2014/416530.

Chen, Na, Zeshui Xu, & Meimei Xia (2013). "Interval-Valued Hesitant Preference Relations and Their Applications to Group Decision Making." *Knowledge-Based Systems*, 37(January), 528–540. https://doi.org/10.1016/j.knosys.2012.09.009.

Chen, Zhen-Song, Yi Yang, Xian-Jia Wang, Kwai-Sang Chin, & Kwok-Leung Tsui (2019). "Fostering Linguistic Decision-Making under Uncertainty: A Proportional Interval Type-2 Hesitant Fuzzy TOPSIS Approach Based on Hamacher Aggregation Operators and Andness Optimization Models." *Information sciences*, 500(October), 229–258. https://doi.org/10.1016/j.ins.2019.05.074.

De, Avijit, Sujit Das, & Samarjit Kar (2019). "Multiple Attribute Decision Making Based on Probabilistic Interval-Valued Intuitionistic Hesitant Fuzzy Set and Extended TOPSIS Method." *Journal of Intelligent & Fuzzy Systems*, 37(4), 5229–5248. https://doi.org/10.3233/JIFS-190205.

Deli, Irfan (2020). "A TOPSIS Method by Using Generalized Trapezoidal Hesitant Fuzzy Numbers and Application to a Robot Selection Problem." *Journal of Intelligent & Fuzzy Systems*, 38(1), 779–793. https://doi.org/10.3233/JIFS-179448.

Dorfeshan, Y., & S. Meysam Mousavi (2019). "A Group TOPSIS-COPRAS Methodology with Pythagorean Fuzzy Sets Considering Weights of Experts for Project Critical Path Problem." *Journal of Intelligent & Fuzzy Systems* 36(2), 1375–1387. https://doi.org/10.3233/JIFS-172252.

El Alaoui, Mohamed (2018). "SMART Grid Evaluation Using Fuzzy Numbers and TOPSIS." *IOP Conference Series: Materials Science and Engineering*, 353(1): 012019. https://doi.org/10.1088/1757-899X/353/1/012019.

El Alaoui, Mohamed, & Hussain Ben-azza (2017). "Generalization of the Weighted Product Aggregation Applied to Data Fusion of Intuitionistic Fuzzy Quantities." In *2017 Intelligent Systems and Computer Vision (ISCV)*, 1–6. https://doi.org/10.1109/ISACV.2017.8054908.

El Alaoui, Mohamed, Hussain Ben-azza, & Azeddine Zahi (2016). "New Multi-Criteria Decision-Making Based on Fuzzy Similarity, Distance and Ranking." In *Proceedings of the Third International Afro-European Conference for Industrial Advancement — AECIA 2016*, 138–148. Advances in Intelligent Systems and Computing. Cham: Springer. https://doi.org/10.1007/978-3-319-60834-1_15.

El Alaoui, Mohamed, Khalid El Yassini, & Hussain Ben-azza (2019). "Type 2 Fuzzy TOPSIS for Agriculture MCDM Problems." *International Journal of Sustainable Agricultural Management and Informatics*, 5(2/3), 112–130. https://doi.org/10.1504/IJSAMI.2019.101672.

El-Hefenawy, Nancy, Mohamed A. Metwally, Zenat M. Ahmed, & Ibrahim M. El-Henawy (2016). "A Review on the Applications of Neutrosophic Sets." American Scientific Publishers. January 2016. https://doi.org/10.1166/jctn.2016.4796.

Estrella, Francisco J., Sezi Cevik Onar, Rosa M. Rodríguez, Basar Oztaysi, Luis Martínez, & Cengiz Kahraman (2017). "Selecting Firms in University Technoparks: A Hesitant Linguistic Fuzzy TOPSIS Model for Heterogeneous Contexts." *Journal of Intelligent & Fuzzy Systems*, 33(2), 1155–1172. https://doi.org/10.3233/JIFS-16727.

Fahmi, Aliya, Muhammad Aslam, & Saleem Abdullah (2019). "Analysis of Migraine in Mutlicellular Organism Based on Trapezoidal Neutrosophic Cubic Hesitant Fuzzy

TOPSIS Method." *International Journal of Biomathematics*, *12*(08), 1950084. https://doi.org/10.1142/S1793524519500840.

Fahmi, Aliya, Muhammad Aslam, Fuad Ali Ahmed Almahdi, & Fazli Amin (2019). "New Type of Cancer Patients Based on Triangular Cubic Hesitant Fuzzy TOPSIS Method." *International Journal of Biomathematics*, *13*(01), 2050002. https://doi.org/10.1142/S1793524520500023.

Farhadinia, Bahram, & Zeshui Xu (2019). *Information Measures for Hesitant Fuzzy Sets and Their Extensions*. Uncertainty and Operations Research. Singapore: Springer. https://doi.org/10.1007/978-981-13-3729-1.

Feng, Liu, Fan Chuan-qiang, & Xie Wei-he (2018). "Type-2 Hesitant Fuzzy Sets." *Fuzzy Information and Engineering*, *10*(2), 249–259. https://doi.org/10.1080/16168658.2018.1517977.

Han, Li, Song, Zhang, & Wang (2019). "A New Method for MAGDM Based on Improved TOPSIS and a Novel Pythagorean Fuzzy Soft Entropy." *Symmetry*, *11*(7), 905. https://doi.org/10.3390/sym11070905.

Han, Qi, Weimin Li, Yanli Lu, Mingfa Zheng, Wen Quan, & Yafei Song (2020). "TOPSIS Method Based on Novel Entropy and Distance Measure for Linguistic Pythagorean Fuzzy Sets With Their Application in Multiple Attribute Decision Making." *IEEE Access*, *8*, 14401–14412. https://doi.org/10.1109/ACCESS.2019.2963261.

Hashmi, Masooma Raza, & Muhammad Riaz (2020). "A Novel Approach to Censuses Process by Using Pythagorean M-Polar Fuzzy Dombi's Aggregation Operators." *Journal of Intelligent & Fuzzy Systems*, *38*(2), 1977–1995. https://doi.org/10.3233/JIFS-190613.

Ho, Lun-Hui, Yu-Li Lin, & Ting-Yu Chen (2019). "A Pearson-like Correlation-Based TOPSIS Method with Interval-Valued Pythagorean Fuzzy Uncertainty and Its Application to Multiple Criteria Decision Analysis of Stroke Rehabilitation Treatments." *Neural Computing and Applications*, June. https://doi.org/10.1007/s00521-019-04304-8.

Hsu, Hsi-Mei, & Chen-Tung Chen (1996). "Aggregation of Fuzzy Opinions under Group Decision Making." *Fuzzy Sets and Systems*, *79*(3), 279–285. https://doi.org/10.1016/0165-0114(95)00185-9.

Hussian, Zahid, & Miin-Shen Yang (2019). "Distance and Similarity Measures of Pythagorean Fuzzy Sets Based on the Hausdorff Metric with Application to Fuzzy TOPSIS." *International Journal of Intelligent Systems*, *34*(10), 2633–2654. https://doi.org/10.1002/int.22169.

J. Chen, Zeng S., & Zhang C. (2018). "An OWA Distance-Based, Single-Valued Neutrosophic Linguistic TOPSIS Approach for Green Supplier Evaluation and Selection in Low-Carbon Supply Chains." *International Journal of Environmental Research and Public Health*, *15*(7). https://doi.org/10.3390/ijerph15071439.

Joshi, Deepa, & Sanjay Kumar (2016). "Interval-Valued Intuitionistic Hesitant Fuzzy Choquet Integral Based TOPSIS Method for Multi-Criteria Group Decision Making." *European Journal of Operational Research*, *248*(1), 183–191. https://doi.org/10.1016/j.ejor.2015.06.047.

Joshi, Dheeraj Kumar, & Sanjay Kumar (2018). "Trapezium Cloud TOPSIS Method with Interval-Valued Intuitionistic Hesitant Fuzzy Linguistic Information." *Granular Computing*, *3*(2), 139–152. https://doi.org/10.1007/s41066-017-0062-5.

Junaid, Muhammad, Ye Xue, Muzzammil Wasim Syed, Ji Zu Li, & Muhammad Ziaullah (2020). "A Neutrosophic AHP and TOPSIS Framework for Supply Chain Risk Assessment in Automotive Industry of Pakistan." *Sustainability*, *12*(1), 154. https://doi.org/10.3390/su12010154.

Karaaslan, Faruk, & Fatih Hunu (2020). "Type-2 Single-Valued Neutrosophic Sets and Their Applications in Multi-Criteria Group Decision Making Based on TOPSIS Method." *Journal of Ambient Intelligence and Humanized Computing*, January. https://doi.org/10.1007/s12652-020-01686-9.

Karaşan, Ali, Eda Bolturk, and Cengiz Kahraman (2020). "An Integrated Interval-Valued Neutrosophic AHP and TOPSIS Methodology for Sustainable Cities' Challenges." In *Intelligent and Fuzzy Techniques in Big Data Analytics and Decision Making*, edited by Cengiz Kahraman, Selcuk Cebi, Sezi Cevik Onar, Basar Oztaysi, A. Cagri Tolga, and Irem Ucal Sari, (pp. 653–661). Advances in Intelligent Systems and Computing. Cham: Springer International Publishing. https://doi.org/10.1007/978-3-030-23756-1_79.

Karaşan, Ali, & İhsan Kaya (2020). "Neutrosophic TOPSIS Method for Technology Evaluation of Unmanned Aerial Vehicles (UAVs)." In *Intelligent and Fuzzy Techniques in Big Data Analytics and Decision Making*, edited by Cengiz Kahraman, Selcuk Cebi, Sezi Cevik Onar, Basar Oztaysi, A. Cagri Tolga, and Irem Ucal Sari, (pp. 665–673). Advances in Intelligent Systems and Computing. Cham: Springer International Publishing. https://doi.org/10.1007/978-3-030-23756-1_80.

Khan, Mohsin, Le Hoang Son, Mumtaz Ali, Hoang Thi Minh Chau, Nguyen Thi Nhu Na, & Florentin Smarandache (2018). "Systematic Review of Decision Making Algorithms in Extended Neutrosophic Sets." *Symmetry*, 10(8), 314. https://doi.org/10.3390/sym10080314.

Khan, Muhammad Sajjad Ali, Faisal Khan, Joseph Lemley, Saleem Abdullah, & Fawad Hussain (2020). "Extended Topsis Method Based on Pythagorean Cubic Fuzzy Multi-Criteria Decision Making with Incomplete Weight Information." *Journal of Intelligent & Fuzzy Systems*, 38(2), 2285–2296. https://doi.org/10.3233/JIFS-191089.

Kilic, Huseyin Selcuk, & Ahmet Selcuk Yalcin (2020). "Comparison of Municipalities Considering Environmental Sustainability via Neutrosophic DEMATEL Based TOPSIS." *Socio-Economic Planning Sciences*, March, 100827. https://doi.org/10.1016/j.seps.2020.100827.

Li, Huimin, Limin Su, Yongchao Cao, & Lelin Lv (2019). "A Pythagorean Fuzzy TOPSIS Method Based on Similarity Measure and Its Application to Project Delivery System Selection." *Journal of Intelligent & Fuzzy Systems* 37(5), 7059–7071. https://doi.org/10.3233/JIFS-181690.

Liang, Decui, & Zeshui Xu (2017). "The New Extension of TOPSIS Method for Multiple Criteria Decision Making with Hesitant Pythagorean Fuzzy Sets." *Applied Soft Computing*, 60(November), 167–179. https://doi.org/10.1016/j.asoc.2017.06.034.

Liang, Decui, Zeshui Xu, Dun Liu, & Yao Wu (2018). "Method for Three-Way Decisions Using Ideal TOPSIS Solutions at Pythagorean Fuzzy Information." *Information sciences*, 435(April), 282–295. https://doi.org/10.1016/j.ins.2018.01.015.

Liang, Weizhang, Guoyan Zhao, & Hao Wu (2017). "Evaluating Investment Risks of Metallic Mines Using an Extended TOPSIS Method with Linguistic Neutrosophic Numbers." *Symmetry*, 9(8), 149. https://doi.org/10.3390/sym9080149.

Liao, Huchang, Ming Tang, Rui Qin, Xiaomei Mi, Abdulrahman Altalhi, Saleh Alshomrani, & Francisco Herrera (2020). "Overview of Hesitant Linguistic Preference Relations for Representing Cognitive Complex Information: Where We Stand and What Is Next." *Cognitive Computation* 12(1), 25–48. https://doi.org/10.1007/s12559-019-09681-9.

Liao, Huchang, & Zeshui Xu (2017). *Hesitant Fuzzy Decision Making Methodologies and Applications*. Uncertainty and Operations Research. Singapore: Springer. https://doi.org/10.1007/978-981-10-3265-3.

Liao, Huchang, Zeshui Xu, Enrique Herrera-Viedma, & Francisco Herrera (2018). "Hesitant Fuzzy Linguistic Term Set and Its Application in Decision Making: A State-of-the-Art Survey." *International Journal of Fuzzy Systems* 20(7), 2084–2110. https://doi.org/10.1007/s40815-017-0432-9.

Lin, Mingwei, Chao Huang, & Zeshui Xu (2019). "TOPSIS Method Based on Correlation Coefficient and Entropy Measure for Linguistic Pythagorean Fuzzy Sets and Its Application to Multiple Attribute Decision Making." *Complexity*, 2019(October), 1–16. https://doi.org/10.1155/2019/6967390.

Liu, Lijia, & Yan Chen (2017) "Interval-Valued Intuitionistic Hesitant Fuzzy Quasi-Choquet Geometric Operators Based TOPSIS Method for Multi-Criteria Group Decision Making." In *2017 29th Chinese Control And Decision Conference (CCDC)*, 2374–2380. https://doi.org/10.1109/CCDC.2017.7978912.

Nabeeh, Nada A., Florentin Smarandache, Mohamed Abdel-Basset, Haitham A. El-Ghareeb, & Ahmed Aboelfetouh (2019). "An Integrated Neutrosophic-TOPSIS Approach and Its Application to Personnel Selection: A New Trend in Brain Processing and Analysis." *IEEE Access*, 7, 29734–29744. https://doi.org/10.1109/ACCESS.2019.2899841.

Naeem, Khalid, Muhammad Riaz, & Deeba Afzal (2019). "Pythagorean M-Polar Fuzzy Sets and TOPSIS Method for the Selection of Advertisement Mode." *Journal of Intelligent & Fuzzy Systems*, 37(6), 8441–8458. https://doi.org/10.3233/JIFS-191087.

Naeem, Khalid, Muhammad Riaz, Xindong Peng, & Deeba Afzal (2019). "Pythagorean Fuzzy Soft MCGDM Methods Based on TOPSIS, VIKOR and Aggregation Operators." *Journal of Intelligent & Fuzzy Systems*, 37(5), 6937–6957. https://doi.org/10.3233/JIFS-190905.

Nancy, & Harish Garg (2019). "A Novel Divergence Measure and Its Based TOPSIS Method for Multi Criteria Decision-Making under Single-Valued Neutrosophic Environment." *Journal of Intelligent & Fuzzy Systems*, 36(1), 101–115. https://doi.org/10.3233/JIFS-18040.

Onar, Sezi Cevik, Başar Oztaysi, & Cengiz Kahraman (2014). "Strategic Decision Selection Using Hesitant Fuzzy TOPSIS and Interval Type-2 Fuzzy AHP: A Case Study." *International Journal of Computational Intelligence Systems*, 7(5), 1002–1021. https://doi.org/10.1080/18756891.2014.964011.

Otay, İrem, & Cengiz Kahraman (2018). "Six Sigma Project Selection Using Interval Neutrosophic TOPSIS." In *Advances in Fuzzy Logic and Technology 2017*, edited by Janusz Kacprzyk, Eulalia Szmidt, Slawomir Zadrożny, Krassimir T. Atanassov, and Maciej Krawczak, (pp. 83–93). Advances in Intelligent Systems and Computing. Cham: Springer International Publishing. https://doi.org/10.1007/978-3-319-66827-7_8.

Otay, İrem, & Cengiz Kahraman (2019). "A State-of-the-Art Review of Neutrosophic Sets and Theory." In *Fuzzy Multi-Criteria Decision-Making Using Neutrosophic Sets*, edited by Cengiz Kahraman and İrem Otay, (pp. 3–24). Studies in Fuzziness and Soft Computing. Cham: Springer International Publishing. https://doi.org/10.1007/978-3-030-00045-5_1.

Oz, Nurdan Ece, Suleyman Mete, Faruk Serin, & Muhammet Gul (2019). "Risk Assessment for Clearing and Grading Process of a Natural Gas Pipeline Project: An Extended TOPSIS Model with Pythagorean Fuzzy Sets for Prioritizing Hazards." *Human and Ecological Risk Assessment: An International Journal*, 25(6), 1615–1632. https://doi.org/10.1080/10807039.2018.1495057.

Peng, Xindong (2019). "New Multiparametric Similarity Measure and Distance Measure for Interval Neutrosophic Set With IoT Industry Evaluation." *IEEE Access*, 7, 28258–28280. https://doi.org/10.1109/ACCESS.2019.2902148.

Peng, Xindong, & Jingguo Dai (2018). "Approaches to Single-Valued Neutrosophic MADM Based on MABAC, TOPSIS and New Similarity Measure with Score Function." *Neural Computing and Applications*, 29(10), 939–954. https://doi.org/10.1007/s00521-016-2607-y.

Peng, Xindong, & Jingguo Dai (2020). "A Bibliometric Analysis of Neutrosophic Set: Two Decades Review from 1998 to 2017." *Artificial Intelligence Review*, 53(1), 199–255. https://doi.org/10.1007/s10462-018-9652-0.

Pérez-Domínguez, Luis, David Luviano-Cruz, Delia Valles-Rosales, Jésus Israel Hernández, & Manuel Iván Rodríguez Borbón (2019). "Hesitant Fuzzy Linguistic Term and TOPSIS to Assess Lean Performance." *Applied Sciences*, 9(5), 873. https://doi.org/10.3390/app9050873.

Rani, Pratibha, Arunodaya Raj Mishra, Ghasem Rezaei, Huchang Liao, & Abbas Mardani (2020). "Extended Pythagorean Fuzzy TOPSIS Method Based on Similarity Measure for Sustainable Recycling Partner Selection." *International Journal of Fuzzy Systems*, 22(2), 735–747. https://doi.org/10.1007/s40815-019-00689-9.

Riaz, Muhammad, & Masooma Raza Hashmi (2019). "Soft Rough Pythagorean M-Polar Fuzzy Sets and Pythagorean m-Polar Fuzzy Soft Rough Sets with Application to Decision-Making." *Computational and Applied Mathematics*, 39(1), 16. https://doi.org/10.1007/s40314-019-0989-z.

Rodríguez, Rosa M., Luis Martínez, Francisco Herrera, & Vicenç Torra (2016). "A Review of Hesitant Fuzzy Sets: Quantitative and Qualitative Extensions." In *Fuzzy Logic in Its 50th Year: New Developments, Directions and Challenges*, edited by Cengiz Kahraman, Uzay Kaymak, and Adnan Yazici, (pp. 109–128). Studies in Fuzziness and Soft Computing. Cham: Springer International Publishing. https://doi.org/10.1007/978-3-319-31093-0_5.

Rodríguez, Rosa M., Luis Martínez, Vicenç Torra, Z. S. Xu, & Francisco Herrera (2014). "Hesitant Fuzzy Sets: State of the Art and Future Directions." *International Journal of Intelligent Systems*, 29(6), 495–524.

Roghanian, E., J. Rahimi, & A. Ansari (2010). "Comparison of First Aggregation and Last Aggregation in Fuzzy Group TOPSIS." *Applied Mathematical Modelling*, 34(12), 3754–3766. https://doi.org/10.1016/j.apm.2010.02.039.

Sajjad Ali Khan, Muhammad, Saleem Abdullah, Muhammad Yousaf Ali, Iqtadar Hussain, & Muhammad Farooq (2018). "Extension of TOPSIS Method Base on Choquet Integral under Interval-Valued Pythagorean Fuzzy Environment." *Journal of Intelligent & Fuzzy Systems*, 34(1), 267–282. https://doi.org/10.3233/JIFS-171164.

Sajjad Ali Khan, Muhammad, Asad Ali, Saleem Abdullah, Fazli Amin, & Fawad Hussain (2018). "New Extension of TOPSIS Method Based on Pythagorean Hesitant Fuzzy Sets with Incomplete Weight Information." *Journal of Intelligent & Fuzzy Systems*, 35(5), 5435–5448. https://doi.org/10.3233/JIFS-171190.

Selvachandran, Ganeshsree, Shio Gai Quek, Florentin Smarandache, & Said Broumi (2018). "An Extended Technique for Order Preference by Similarity to an Ideal Solution (TOPSIS) with Maximizing Deviation Method Based on Integrated Weight Measure for Single-Valued Neutrosophic Sets." *Symmetry*, 10(7), 236. https://doi.org/10.3390/sym10070236.

Senvar, Ozlem, Irem Otay, & Eda Bolturk (2016). "Hospital Site Selection via Hesitant Fuzzy TOPSIS." *IFAC-PapersOnLine*, 8th IFAC Conference on Manufacturing Modelling, Management and Control MIM 2016, 49(12): 1140–1145. https://doi.org/10.1016/j.ifacol.2016.07.656.

Sharma, Himanshu, Abhishek Tandon, P. K. Kapur, & Anu G. Aggarwal (2019). "Ranking Hotels Using Aspect Ratings Based Sentiment Classification and Interval-Valued Neutrosophic TOPSIS." *International Journal of System Assurance Engineering and Management*, 10(5), 973–983. https://doi.org/10.1007/s13198-019-00827-4.

Smarandache, Florentin (1999). *A Unifying Field in Logics. Neutrosophy: Neutrosophic Probability, Set and Logic*. American Research Press, Rehoboth.

Tavakkoli-Moghaddam, Reza, Hossein Gitinavard, Seyed Meysam Mousavi, & Ali Siadat (2015). "An Interval-Valued Hesitant Fuzzy TOPSIS Method to Determine the Criteria Weights." In *Outlooks and Insights on Group Decision and Negotiation*, edited by Bogumił Kamiński, Gregory E. Kersten, and Tomasz Szapiro, (pp. 157–169). Lecture Notes in Business Information Processing. Cham: Springer International Publishing. https://doi.org/10.1007/978-3-319-19515-5_13.

Tehrim, Syeda Tayyba, & Muhammad Riaz (2019). "A Novel Extension of TOPSIS to MCGDM with Bipolar Neutrosophic Soft Topology." *Journal of Intelligent & Fuzzy Systems*, 37(4), 5531–5549. https://doi.org/10.3233/JIFS-190668.

Torra, Vicenç (2010). "Hesitant Fuzzy Sets." *International Journal of Intelligent Systems*, 25(6), 529–539. https://doi.org/10.1002/int.20418.

Wang, Haibin, Florentin Smarandache, Rajshekhar Sunderraman, & Yan-Qing Zhang (2005). *Interval Neutrosophic Sets and Logic: Theory and Applications in Computing: Theory and Applications in Computing*. Infinite Study.

Wang, Lina, Hai Wang, Zeshui Xu, & Zhiliang Ren (2019). "The Interval-Valued Hesitant Pythagorean Fuzzy Set and Its Applications with Extended TOPSIS and Choquet Integral-Based Method." *International Journal of Intelligent Systems*, 34(6), 1063–1085. https://doi.org/10.1002/int.22086.

Wu, Jian, Xiao-Di Liu, Zeng-Wen Wang, & Shi-Tao Zhang (2019). "Dynamic Emergency Decision-Making Method With Probabilistic Hesitant Fuzzy Information Based on GM (1,1) and TOPSIS." *IEEE Access*, 7, 7054–7066. https://doi.org/10.1109/ACCESS.2018.2890110.

Wu, Zhibin, Xue Chen, & Jiuping Xu (2017). "TOPSIS-Based Approach for Hesitant Fuzzy Linguistic Term Sets with Possibility Distribution Information." In *2017 29th Chinese Control And Decision Conference (CCDC)*, 7268–7273. Chongqing, China: IEEE. https://doi.org/10.1109/CCDC.2017.7978497.

Wu, Zhibin, Jiuping Xu, Xianglan Jiang, & Lin Zhong (2019). "Two MAGDM Models Based on Hesitant Fuzzy Linguistic Term Sets with Possibility Distributions: VIKOR and TOPSIS." *Information sciences*, 473(January), 101–120. https://doi.org/10.1016/j.ins.2018.09.038.

Xian, Sidong, & Hailin Guo (2020). "Novel Supplier Grading Approach Based on Interval Probability Hesitant Fuzzy Linguistic TOPSIS." *Engineering Applications of Artificial Intelligence*, 87(January), 103299. https://doi.org/10.1016/j.engappai.2019.103299.

Xu, Zeshui (2014). *Hesitant Fuzzy Sets Theory*. Studies in Fuzziness and Soft Computing. Springer International Publishing. https://doi.org/10.1007/978-3-319-04711-9.

Yager, Ronald R. (2013). "Pythagorean Fuzzy Subsets." In *2013 Joint IFSA World Congress and NAFIPS Annual Meeting (IFSA/NAFIPS)*, 57–61. https://doi.org/10.1109/IFSA-NAFIPS.2013.6608375.

Yager, Ronald R. (2014). "Pythagorean Membership Grades in Multicriteria Decision Making." *IEEE Transactions on Fuzzy Systems*, 22(4), 958–965. https://doi.org/10.1109/TFUZZ.2013.2278989.

Yang, Wei, & Yongfeng Pang (2018). "New Multiple Attribute Decision Making Method Based on DEMATEL and TOPSIS for Multi-Valued Interval Neutrosophic Sets." *Symmetry*, 10(4), 115. https://doi.org/10.3390/sym10040115.

Yang, Yi, Heng Ding, Zhen-Song Chen, and Yan-Lai Li (2016). "A Note on Extension of TOPSIS to Multiple Criteria Decision Making with Pythagorean Fuzzy Sets." *International Journal of Intelligent Systems*, 31(1), 68–72. https://doi.org/10.1002/int.21745.

Yao, Yiyu (2012). "An Outline of a Theory of Three-Way Decisions." In *Rough Sets and Current Trends in Computing*, edited by JingTao Yao, Yan Yang, Roman Słowiński, Salvatore Greco, Huaxiong Li, Sushmita Mitra, and Lech Polkowski, 1–17. Lecture Notes in Computer Science. Berlin, Heidelberg: Springer. https://doi.org/10.1007/978-3-642-32115-3_1.

Yildiz, Aslihan, Ertugrul Ayyildiz, Alev Taskin Gumus, & Coskun Ozkan (2020). "A Modified Balanced Scorecard Based Hybrid Pythagorean Fuzzy AHP-Topsis Methodology for ATM Site Selection Problem." *International Journal of Information Technology & Decision Making*, January, 1–20. https://doi.org/10.1142/S0219622020500017.

Yu, Chunxia, Yifan Shao, Kai Wang, & Luping Zhang (2019). "A Group Decision Making Sustainable Supplier Selection Approach Using Extended TOPSIS under Interval-Valued Pythagorean Fuzzy Environment." *Expert Systems with Applications*, 121(May), 1–17. https://doi.org/10.1016/j.eswa.2018.12.010.

Yucesan, Melih, & Muhammet Gul (2020). "Hospital Service Quality Evaluation: An Integrated Model Based on Pythagorean Fuzzy AHP and Fuzzy TOPSIS." *Soft Computing*, 24(5), 3237–3255. https://doi.org/10.1007/s00500-019-04084-2.

Zeng, Shouzhen, Dandan Luo, Chonghui Zhang, & Xingsen Li. (2020). "A Correlation-Based TOPSIS Method for Multiple Attribute Decision Making with Single-Valued Neutrosophic Information." *International Journal of Information Technology & Decision Making*, *19*(01), 343–358. https://doi.org/10.1142/S0219622019500512.

Zhang, Qinghua, Qin Xie, & Guoyin Wang (2016). "A Survey on Rough Set Theory and Its Applications." *CAAI Transactions on Intelligence Technology*, *1*(4), 323–333. https://doi.org/10.1016/j.trit.2016.11.001.

Zhang, Xiaolu (2016). "A Novel Approach Based on Similarity Measure for Pythagorean Fuzzy Multiple Criteria Group Decision Making." *International Journal of Intelligent Systems*, *31*(6), 593–611. https://doi.org/10.1002/int.21796.

Zhang, Xiaolu, & Zeshui Xu (2014). "Extension of TOPSIS to Multiple Criteria Decision Making with Pythagorean Fuzzy Sets." *International Journal of Intelligent Systems*, *29*(12), 1061–1078. https://doi.org/10.1002/int.21676.

Zhang, Xiaolu, & Zeshui Xu (2017). *Hesitant Fuzzy Methods for Multiple Criteria Decision Analysis.* Studies in Fuzziness and Soft Computing 345. Switzerland: Springer International Publishing. https://doi.org/10.1007/978-3-319-42001-1.

Zhang, Zhiming (2017). "Hesitant Fuzzy Linguistic TOPSIS Method Using a Possibility-Based Comparison Approach for Multi-Criteria Decision-Making with Hesitant Fuzzy Linguistic Term Sets." *Journal of Intelligent & Fuzzy Systems*, *33*(6), 3309–3322. https://doi.org/10.3233/JIFS-161971.

Zhou, Wei, & Zeshui Xu (2020). *Qualitative Investment Decision-Making Methods under Hesitant Fuzzy Environments.* Studies in Fuzziness and Soft Computing 376. Switzerland: Springer International Publishing. https://doi.org/10.1007/978-3-030-11349-0.

Index

Note: Page numbers in **bold** represent tables, page numbers in *italics* represent figures.

aggregation, vii, ix, xii, 44, 107–109, 111, 127, 135–138, 140, 142–143, **143**, 150, **151**, 167, 179, 188–189
agreement, 111, 176, 183
algorithm, vii, 32, 74, 76, 89, 95, 99–100, 104, 107, 109, 112–115, 118–119, 122, 125, 127, 135, 137, 142–143, 150, 155, 157–160, 167, 176–179, 182, 184, 187
alternative(s), vii, xi–xii, 2, 41–44, 46, **46**, 47, **47**, 48, **48**, 49, 53, **53**, 54, **55**, 56, **56**, 57, 65–67, **67**, 68, **68**, 69–70, **70**, 71, **72**, 73–74, **74–75**, 76, **77–78**, 79, **81–83**, **85–88**, 89, 95, 97, **98–99**, 100, **100–101**, 102–103, **103**, 104, **104**, 105, **105**, 106–107, 110–111, 113–114, **114**, 115, **115**, 116, **117–118**, 119, **119–121**, 122, **122**, 123–124, **124**, 125, **125**, 126, **126**, 127, 136–137, 143, 150, **153**, 154–155, **155–156**, 157, **157**, 158, **158–159**, 160, **161**, 162, **162–163**, 164–165, **165**, 166, **166–167**, 178–179, **179–181**, 184–185, **186**, 187, **187**, 188, **188–189**
ambiguity, 1–3, **5**, 7, 9
Analytic Hierarchy Process/AHP, vii, 42, 44, **45**, 46, 58, 69, 71, 95, 177, 189
ANP, 71, 183
application(s), 21, 23, 24, 35, 58, 111, 117, 136, 177, 183, 189
approach(es), iii, viii, xii, 1–2, 4, 18, 22, 37, 42, 44, 65, 71, 74, 76, 79, 89, 95, 98, **98–100**, 104, **104–105**, 108, 111–113, 117, 127, 136, 141, 150, 157, 164, 167, 175, 177, 179, 183, 188–189
approximation(s), 1, 4
Aristotle, 1, 16, 25
arithmetic, 7, 35, 98
axiom(s), 3–4

background(s), xi–xii, 37, 111, 135
binary, 7, 16
bipolar, 177, 183
bivalence, 16
Bochvar, 18, **20–21**, 25
Boolean, 15–16, 25, 31

characteristic function, *16*, 31, *32*
classical logic, xi, 15–16, *16*, 25, 37
closeness coefficient(s), 66, **68**, 69, 73, 76, 97–99, **100**, **105**, 107, 114, 116, **122**, **126**, 143, **158**, *163*, 165, 177–179, **181**, 185, 188, **189**
COmplex PRoportional ASsessment/COPRAS, 73, 177
complexity, 4, 6, 32, 42, 47
consensus/consensual, vii–ix, xii–xiii, 46, 84, 108, 110–111, *111*, 112–113, **113–114**, 115–116, **116–117**, 118, **124**, 127, 135, 141–143, **156**, 157–158, **162**, 167, 176, 189
convergence, 143, 167
correlation, 73, 79, **89**, **151**
criterion/criteria, vii, xi, 8, 41–46, **46**, 47, **47**, 48–50, **51**, 53, **53–54**, 56–58, 65–67, **67**, 68, **68**, 69–70, **70**, 71, **72**, 73–74, **74**, **75**, 76, 79, **80**, 84, 89, 95–97, **98**, 99, **99**, 100, **100**, 101, **101**, 102–103, **103**, 104, **104**, 105, **105**, 106–110, 113–115, **115–118**, 119, **119–121**, 122–124, **124**, 125, **125–126**, 136–137, 150, 153, **153**, 154–155, **155**, 156, **156–157**, 158–159, **159**, 160, **161–162**, 163–165, **165**, 166, 177–179, **179–181**, 184–185, **186**, 187, **187**, 188, **188**

decision(s)/decision maker(s)/decision-making, v, vii, xi–xii, 5, 41–44, 46, 50, 58, 65–66, **67**, 68, 69–71, 79, 95–98, **98**, 99, **99**, 100, **100**, 101–103, **103–104**, 105–108, 110,

199

112, **113**, 114–115, **115**, 119, **120**, 122, 124, **124**, 125, **125–126**, 135–137, 139, 142–143, 150, **151–152**, 154–155, **155–156**, 158, 160, **161–162**, 163–165, 167, 176–179, **179**, 182–185, **186**, 187, **187**, 188, **188**
DEMATEL, 184
distance(s), vii, xii, 50, 53, **54**, 66, **68**, 69, 76, **77**, 84, **85–88**, 89, 97–98, **100**, 102, 104, **105**, 106, 109, **110**, 111, 114, **114**, 115, 118, **122**, 123, 126, **126**, 127, 135, 140–143, 150, **151–152**, 154, 157, **158**, 159, 160, 162–163, **163**, 164–166, **166**, 167, 175–179, **181**, 182–183, 185, 188, **189**
disagreement, 42
duality, 8, 23–25, 189

Einstein, 4, 15, 24
ELimination Et Choix Traduisant la REalité/ Elimination and Choice Translating Reality/ ELECTRE, vii, 46, **47**, 49, **49**, 50, 71, 73, 95, 183
ELECTRE I, 49, 73
ELECTRE II, 49
ELECTRE III, 49
ELECTRE IV, 49
ELECTRE IS, 49
ELECTRE TRI, 49
entropy, 150, **152**, 163, 177
Eubulides, 1
exact sciences, 3–4

function(s), viii, xii, *16*, 18, 31–32, *32*, 34–37, 41, 50, 51, **51**, **55**, 79, 96, 108, *108*, 111–112, 135, 137–138, 143, **143**, 150, **151**, 154, 167, 177, 180–182
fuzzy/fuzziness, i, iii, vi–vii, xi–xiii, 3–4, 8–9, 16, 23, 25, 31–38, 41, 44, 49, **49**, 53, 76, 95–96, *96*, 97–100, **101**, 104, **104**, 108–109, **110**, 111–113, **113**, **114**, 116–118, **125**, 127, 135, *136*, 138–143, 150, **151**, 153, **153**, 154, 157–158, **158**, 159, **159**, 163–165, 167, 175–179, **179–180**, 181, **181**, 182, *184*, **187**, 188–189

GRA, 157
group, xii–xiii, 50, 57, 74, **104**, 106–107, 110, **111**, 114–115, 135, **151–152**, 178–179, **180–181**, 184, 187, **187–188**

Heisenberg, Werner, 5, 24
hesitancy, 36, 141
hesitant, vii, xii, 150, 164, 184, *184*, 188–189
Huygens, 23

ideal solution(s), vii, xii–xiii, 54, 65–66, **68**, 69, 73–74, **74–75**, 76, **78**, 79–80, **81–83**, 84, 89, 96–98, **100**, 101–104, **105**, 106–107, 109, 113–114, **114**, 115–116, 122–123, 125–126, **126**, 127, 153–154, 156–157, **158**, 159–163, **163**, 164–166, **166**, 167, 178–179, **181**, 184–185, 187–188, **189**
imprecision, 4
importance, 7, 49, 100, **101**, 103, 114, 122, 124, 150, **153**, 155, **156**, 158, **159**, 160, **161**, 178–179, 184, **186**, 187
incompleteness, 5, **5**, 9
indeterminacy/indeterminate, 1, 5, **5**, 6, 9, 17, 175, 180–181, 188
indicator, 79
information, vii, xi–xii, 5, **5**, 6–9, 42, 46, 116, 135, 150, 160, 163, 167, 177, 179
integration, vii, xii, 135, 167
interpretation, 1, 4, 22, 24, 25, 44, 74, 76, 107, 163, 164
interval-valued, vii, xii, 35–36, 108, *108*, 109, 139–143, 150, **151**, 153, **153**, 154, 157, 159, **159**, 177, 181–182, 189
intuitionistic, v, vii–viii, xii, 8, 24, 36–37, 135, *136*, 138–141, 143, 150, **151**, 153, **153**, 154, 157–159, **159**, 163–166, **166**, 167, 175–176, *176*, 177, 179, 181–182, 189
iteration(s), 42, 112, 127, 142–143, **143–149**

Kleene, S. C, 17–18, **18–19**, 25
knowledge, vii, xi, 1, 5, 74, 76, 108

linguistic variable(s), 79, 95, 100, **101**, 102–105, 114, **115**, 119, 122, 124, **125**, 136, 143, 150, **153**, 155, 158, **159**, 160, **161**, 184–185, **186**, 187
LINMAP, 151
logic(s)/logical, iii, xi, xii, 1–4, 8–9, 15–18, 21–25, 31, 34, 37, 49, 53, 79, 95, 108, 137–139, 157, 177
Łukasiewicz, Jan, 16–18, **17**, **18**, 21, 22, 24–25

management, ix, 3, 6, 43, 113, 154
MATLAB, vii, xii, 67, 89, 104, 127
mathematical, vii, xi–xii, 3–4, 8, 15, 18, 25, 37, 111, 135, **152**, 177, 189
matrix, 44–45, **46**, 47, 65–66, 68, 69, **70**, **75**, 95–98, **99–100**, 101, 103, **104–105**, 106, 108, 114–115, **120**, 122, 125, **125–126**, 153, 156, 159–160, **162**, 165, **165**, 176, 178–179, 183–184, 187, **187–188**
membership function, *16*, 32, *32*, 34–37, *108*, 154
method(s), vii–viii, xii, 2, 4–5, 7, 18, 24, 35, 37, 41–44, 46, 49–50, 56, 58, 65, 70–71, 73, 76, 79–80, 84, 89, 95, 98–99,

Index

108–109, **110**, 112–113, 116, 126–127, 135, 141, 143, 150, 154, 163, 177, 189
metric(s), xii, 54, 56, 79, 84, **85–88**, 89, 142–143, 150, **151**, **152**
model(s), vii, xi, 3–5, **5**, 6, 8–9, 95, 97, 135, 150, 177
multiattribute, 41–43, 46, 69, 95
multicriteria, v, 41–42, 53, 79, 99, 136
multiobjective, 41
multivalent, 21, **22**, 25
multivalued/multiple-valued/many valued, vii, 16, 23, 25
m-polar, 177

necessity, 16
neutrosophic, viii, xii, 179–184, *184*, **186**, 189
Newton, 23
nonclassical logics, v, xi, 15, 23, 25, 31
nonmembership function, 35–36, 154, 177
normalization/normalized, vii, xii, 36, 65–66, **67**, 68–69, 73–74, **75**, 76, **77**, 79–80, **80–83**, 84, 89, 95–98, **99–100**, 101, 103, **104–105**, 106, 108–109, 110, 114–115, **118–120**, 122, 125, **125–126**, 127, 150, **151**, 153, 156, **157**, 159–160, **162**, 163, 178–179, **180–181**, 184, 187, **188**

operation(s), viii, xii, 33–35, 37, 41, 98
operator(s), xii, 37, **110**, 135, 137–140, 142–143, **143**, 150, **151–152**, 175–176, 182, 188–189
optimal/optimized/optimization, vii, ix, xii, 23, 41, 54, 74, 76, 112, 143, **143**, 150, 177
outranking, 50, 52, 141

paradox, 1, 18
particle, 23–25
Peirce, 1–2
performance(s), 43–44, 71, 84, 137, 160, 189
philosophy/philosophical, 3, 15–16, 21, 25, 54
Planck, Max, 2–3
Post, 16, 22, 25
possibility/possibilities/possibilistic, xi, 8, 16–17, **151**
precision, 1–2, 4
probability/probabilities/probabilistic, vii, xi, 3–6, 8, 24, 49, 150, 189
problem(s), xi–xii, 2–4, 6, 18, 25, 41–43, 49–50, 71, 73, 79, 97, 110, 135, 177
proposition, 1, 17
Pythagorean, viii, xii, 112, 164, 175–176, *176*, 177–179, **179–180**, 181, **181**, 184, *184*, 189

qualitative, vii, 2, 95, 97, 113–115, **115–117**, **119**, 178, 184
quantitative, vii, 2, 7, 95, 97,113–115, **116**, **118**
quantum mechanics/quantum physics/quantum theory, vii, xi, 18, 23–25

rank(s)/ranking(s), vii, xii, 37, 41, 44, 46, 49, **49**, 50, 52–53, 57–58, 66, **68**, 69–71, **72**, 73, **74**, 76, **78**, 79–80, **81–83**, 84, **85–88**, 89, **89**, 96–99, **100**, 102, 104, **105**, 107, 109–110, **110**, 113–114, **114**, 116, **122**, 123, 126, **126**, 127, 153–154, 157, **158**, 160, 162–163, **163**, 164–167, **167**, 178–179, **181**, 185, 188, **189**
relation, 73
rough sets, 8, 177

Russell, Bertrand, 4
Schrödinger, 15, 24–25
similarity/dissimilarity, ix, xi–xii, 43, 111–112, 118, 142–143, 150, **151–152**, 157, 159, 177
Simple Additive Weighting/SAW, vii, 58, 71, 73, 95, 116
single-valued, 183
singleton(s), 36
soft set(s), 183
subproblem(s), 71, 73, **74**
syllogism, 16, 25

theory, vii, xi–xii, 8–9, 15, 23, **151**, 177
TOPSIS, i, iii, v, vii–viii, xii–xiii, 37, 41, 54, 58, 65, **67**, 71, **72**, 73, 74, 76, 79, 84, 89, 95–96, *96*, 98–100, 104, 108–110, **110**, 111, *111*, 112–113, **113–114**, 116–117, 127, 135, *136*, 141, 143, **151–152**, 157–158, 167, 175, 177, 179, 183–184, *184*, 189
trivalent logic, 16–18, 21, 24–25

uncertainty, vii, xi, 1–5, **5**, 6–9, *9*, 31, 34, 43, 49, 53

vague/vagueness, 1–3, 7, 9, 31, 43, 73, 136
VIKOR, vii, 53, 57, **57**, 58, 71, 73, 157, 177, 189

wave, 23–25
weight(s), viii, xi–xii, 2, 7, 43, 45, 47, **47**, 50, 53, 57–58, 65, **67**, 68, **72**, 73, 76, 89, 95, **98**, 100, 103, **103**, 104, **104**, 106, 110, 112, 114–115, **116**, 122, 125, 136, 138–139, 142–143, 150, **151–152**, 155–156, 158, 160, **162**, 163–167, 175, 177–179, 182–184, 187, **187**, 188

weighted incoherencies, 112
weighted normalized, 65–66, **67**, 69, 73–74, **75**, 76, **77**, 79, 96–98, **99–100**, 101, 103, **105**, 106, 108, 114–115, **120**, 122, 125, **126**, 153, 156, **157**, 159–160, **162**, 163, 178–179, **181**, 184, 187, **188**

weighted product, 43
Weighted Sum/WS, 43, 46–47, 58

Young, 24

Printed in the United States
by Baker & Taylor Publisher Services

Printed in the United States
by Baker & Taylor Publisher Services